Beneficial Modifications of the Marine Environment

A Symposium sponsored by the
NATIONAL RESEARCH COUNCIL
NATIONAL ACADEMY OF SCIENCES
NATIONAL ACADEMY OF ENGINEERING
and the
U.S. DEPARTMENT OF THE INTERIOR

March 11, 1968
Washington, D.C.

NATIONAL ACADEMY OF SCIENCES
WASHINGTON, D.C.   1972

NOTICE: The symposium at which these papers were read and discussed was sponsored by the National Research Council and held at its Eleventh Annual Meeting in 1968. The participants in the symposium were selected for their scholarly competence and for the particularly relevant specialty that enabled them to contribute to the exposition of the subject under discussion. The participants were responsible for their own papers, and the views expressed were their own.

Available from

Printing and Publishing Office
National Academy of Sciences
2101 Constitution Avenue, NW
Washington, D.C. 20418

ISBN 0-309-02034-4
Library of Congress Catalog Card Number 72-80691

Printed in the United States of America

Preface

The Symposium on Beneficial Modifications of the Marine Environment, cosponsored by the National Research Council and Department of the Interior, was convened as part of the Research Council's Eleventh Annual Meeting, in Washington, D.C., March 10–12, 1968. The agenda of the Symposium was prepared by Dr. S. Fred Singer, Deputy Assistant Secretary for Scientific Programs, in the Office of the Assistant Secretary for Water Quality and Research, U.S. Department of the Interior, and the Symposium was arranged and carried out by the Division of Earth Sciences of the National Research Council.

This proceedings volume has been assembled from materials prepared by the Symposium's speakers and discussants, based on their presentations. Each has subsequently been given an opportunity to review and update his contribution in the light of further developments.

The impetus for this Symposium on Beneficial Modifications of the Marine Environment came from the activities of the Federal Marine Council and the National Marine Commission, both of which were set up under the Marine Resources and Engineering Development Act of 1966. Both the Council and the Commission have a real need for projecting the long-range future of our activities in the marine environment: What kind of projects will be carried out in the ocean in the near future? What kind of projects are likely to be undertaken in the distant future? Will we have the necessary scientific background, manpower, facilities, or organizational structure? What are the economic and political considerations, national and international, of such projects?

All these questions are important. But, for the purposes of this Symposium, we concentrated on the purely scientific and technical aspects of various plans that involve the modification of the marine environment—and here we include the Great Lakes, as is done in the Marine Resources Act.

This does not mean that we should neglect the nontechnical aspects, but it is necessary first to understand the scientific and technical problems involved in dealing with the marine environment. It is also imperative to understand the consequences of any modification of the environment.

PLANETARY ENGINEERING

Modification of the physical environment is nothing new. The land environment has been subject to planned modification since the beginning of man's existence. But it is just now that we have available the technical means for truly large-scale modifications—modifications that can span a good portion of the globe. We are developing the field of planetary engineering, because a large-scale modification of one part of the earth may have far-reaching effects all around the globe. We are approaching lower-cost sources of nuclear energy and, with the help of satellites, we can keep track of any changes on the earth's surface and in the oceans and atmosphere.

OUTLINE OF THE SYMPOSIUM CONTENT

A number of imaginative proposals are put forth in this Symposium, tempered, I trust, by the invited discussants, so that you can form a balanced view of both their scientific and technical feasibility and their desirability. These ideas and proposals must be placed in the public forum and be available for discussion by scientists and technologists before they can be brought into the political decision area.

Two of the proposals involve the interaction between the ocean surface and the atmosphere. The paper by Dr. J. O. Fletcher discusses the reason that the presence or absence of ice on the sea can influence the formation of climate and looks into the following two questions: If the pack ice were removed, would the Arctic Ocean remain an open sea? What are the possibilities for artificially influencing large-scale climate by influencing the extent of ice on the sea? The possibility of doing this may seem distant to us now, but let us imagine that we were faced with such a technical proposal from another country: How would we respond?

The paper by Dr. R. D. Gerard and Dr. J. L. Worzel deals with a more immediate problem—the possibility of extracting appreciable quantities of atmospheric moisture to provide water for the islands in the Caribbean. This is to be accomplished by pumping cold bottom water to the surface of the ocean—a kind of artificially produced upwelling. Here, again, you will note that we may take advantage of the fact that we can interfere with natural processes that are almost completely balanced by introducing a small "push" at the right place and at the right time.

Some of the projects discussed here are not really "far out." Dr. W. C. Ackermann's paper on water transfers into or between the Great Lakes and Dr. D. W. Pritchard's paper on modification of flow in estuaries discuss projects that we may want to undertake in the relatively near future. There is no question in my mind that the population increase and the resulting pressures on natural resources will force us to make increasing use of marine resources and to make many modifications in various areas of the marine environment. Let us hope that they will all be beneficial. By exposing them at an early stage and by giving them careful and complete discussion, we shall ensure the achievement of this goal.

Because modification of the water and marine environment is of special concern to the Department of the Interior, we have asked Secretary Udall to introduce this Symposium.

>S. FRED SINGER, *Chairman*
>Symposium on Beneficial Modifications
>of the Marine Environment

Contents

INTRODUCTION
    Stewart L. Udall   1
ICE ON THE OCEAN AND WORLD CLIMATE
    J. O. Fletcher   4
    DISCUSSION: *William L. Donn* and *David M. Shaw*   49
    DISCUSSION: *Sigmund Fritz*   61
ATMOSPHERIC MOISTURE EXTRACTION OVER THE OCEAN
    Robert D. Gerard and *J. Lamar Worzel*   66
    DISCUSSION: *Helmut E. Landsberg*   78
    DISCUSSION: *Earl G. Droessler*   82
WATER TRANSFERS: POSSIBLE DE-EUTROPHICATION OF THE GREAT LAKES
    William C. Ackermann   85
    DISCUSSION: *J. P. Bruce*   101
MODIFICATION AND MANAGEMENT OF WATER FLOW IN ESTUARIES
    Donald W. Pritchard   104
    DISCUSSION: *L. Eugene Cronin*   112
    DISCUSSION: *Joseph M. Caldwell*   115

Introduction

For openers, I want to toss in a welcome and a warning. First, as Secretary of the department of government primarily concerned with the nation's natural resources, their development, and their proper use, I am delighted that we are cosponsoring this Symposium on Beneficial Modifications of the Marine Environment.

Second, I feel constrained to sound my customary note of caution—something that has become almost instinctive with me since I became familiar with the various conservation problems that seem to exist in splendid isolation within different sectors of our environment.

I have found that most of these seemingly unrelated environmental problems can be the result of some solution to a former problem—in extreme cases a solution arrived at in haste, applied in ignorance, and pregnant with future environmental headaches.

Furthermore, I have found that the problems are seldom as unrelated as those who come to me with them think they are. A long-chain hydrocarbon biocide that can help produce a crop of rosy apples in an Oregon valley can also mean, under certain circumstances, a crop of dead fish on a Washington beach or river bank.

A huge supertanker means faster transport of millions of barrels of oil. Add carelessness or an accident, and the same tanker can mean the frantic shriek and flap of death for a red-eyed loon, a slim-billed merganser, a tiny puffin—all part of a carcass-strewn beach.

The Department of the Interior is responsible for mineral resources, on land and on the continental shelf; for land resources on the public lands, the national parks and, again, the continental shelf; and for living resources on the land and in the lakes and rivers and oceans.

These responsibilities include water resources in all forms. We have programs that are concerned with all the aspects of the hydrological cycle. We support programs designed to increase the precipitation of atmospheric moisture; we concern ourselves with management of water in rivers, in reservoirs, in the ground; we work at water quality in rivers and lakes and estuaries; and we are much involved with all the biota in the waters.

There are ten Interior Department bureaus with well-defined marine interests, but many of our bureaus don't distinguish between land and marine programs. The Geological Survey, for example, extends its studies from the land area of the United States onto the continental shelf in a continuous way, with only the techniques for exploration changing.

The Department of the Interior, then, is the leading civilian agency in the field of marine resources. It is well that this is so, for we also have the major share of land-resource responsibilities, and in this "Age of Ecology" there is no way of separating, in fact, the interrelationships and interworkings of land and ocean. It is best that we do not try to separate too far our attempts to deal with them.

You who are about to concern yourselves with the modification of the marine environment will not, I am sure, overlook the long history of land modification.

Some land modification has been both beneficial and gentle. Most agriculture would come under this classification. Raising food for a rising population would seem to be purely beneficial, but what of indiscriminately applied pesticides that run off fields, produce massive river fish-kills, and even build up in deep-sea tuna and Antarctic penguins? Mining coal for warmth and power sounds beneficial, but what about open-pit and strip methods that ravage and ruin whole regions?

It seems to me that we must be much longer-sighted than we have been in the past—that we must be careful to apply much wider vision when we consider modification of even such a huge consideration as the earth's oceans.

Remember how the pioneers felt about the seemingly endless prairie—and the forests—and the buffalo—and the passenger pigeon? There is little enough left of some of these, and none at all of one.

        STEWART L. UDALL, *Secretary*
        Department of the Interior

J. O. Fletcher*

THE RAND CORPORATION

Ice on the Ocean and World Climate

About 10 percent of the ocean area in the Northern Hemisphere is covered by floating ice in winter (Figure 1); in the Southern Hemisphere, over a larger ocean area, the figure is about 13 percent. The extent of this pack ice varies greatly during the year and from year to year. It has long been observed that these variations show a close correlation with many indices of climatic change. As the arctic ice pack recedes, storm tracks tend to go farther north, and midlatitude rainfall patterns tend to shift eastward.

This paper attempts to explain why the variable extent of ice on the sea is a very sensitive climatic lever that can amplify the effects of small changes in global heating. The prospect for influencing these processes is also discussed.

PHYSICAL EFFECTS OF AN ICE COVER

To understand fully why the presence or absence of ice cover has such an enormous effect on the heat budget of the atmosphere, one must

*Now with the National Science Foundation.

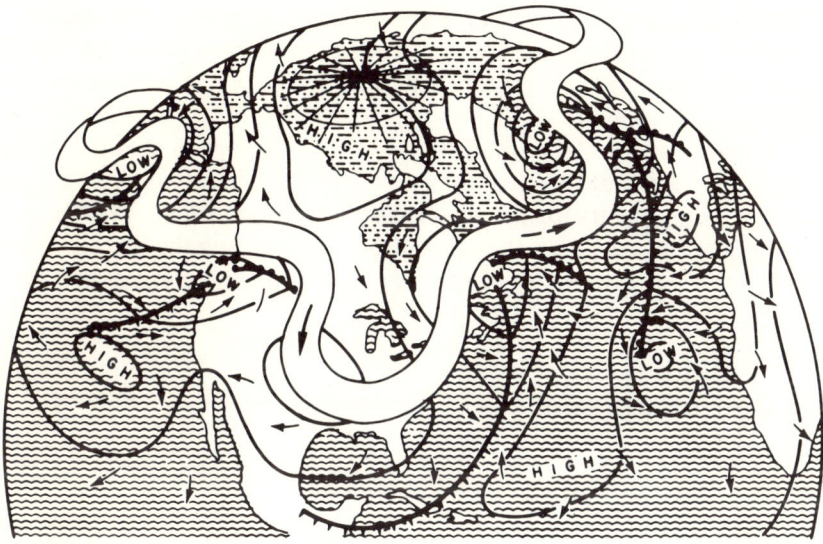

FIGURE 1 Variable extent of Arctic pack ice as related to variable intensity of atmospheric circulation. The general circulation of the atmosphere is forced by gradients arising from the net heat loss to space in polar regions and net heat gain at low latitudes. The intensity of polar cooling is influenced by the extent of ice on the ocean.

look closely at the behavior of all the heat-budget components, but in the simplest terms, the presence of an ice cover effectively prevents heat exchange between the ocean and the atmosphere, both in winter and in summer. For example, in January the mean air temperature at the earth's surface in the central Arctic is about $-30°$ C, while 2 or 3 m below the surface, the temperature of the ocean water is near $-2°$C; yet the ice and its snow cover provide such good insulation that little more than one kcal/cm^2/month reaches the surface from below. The ice surface radiates heat to space, and this heat loss simply cools the ice surface until it is cold enough to drain from the atmosphere the heat needed to balance the loss. The thermal participation of the ocean is greatly suppressed when ice is present. If the ice were not there, most atmospheric heat would be obtained from the relatively warm ocean.

In summer, an open ocean would absorb about 90 percent of the solar radiation reaching the surface, while the pack ice reflects 60 or 70 percent of the incident sunlight. Thus, the presence of the ice suppresses heat loss by the ocean in winter and heat gain by the ocean in summer. For the atmosphere the reciprocal relation applies: Over pack ice, the atmosphere cools more intensely during winter and warms more intensely in summer.

These effects are quantitatively illustrated by Figures 2 and 3, which show the results of detailed estimates from a previous study of the heat-budget components for the Arctic basin (Fletcher, 1965). The "ice-in" curve of Figure 2 shows the amount of heat reaching the surface from below during each month of the year. The heat gained in summer is

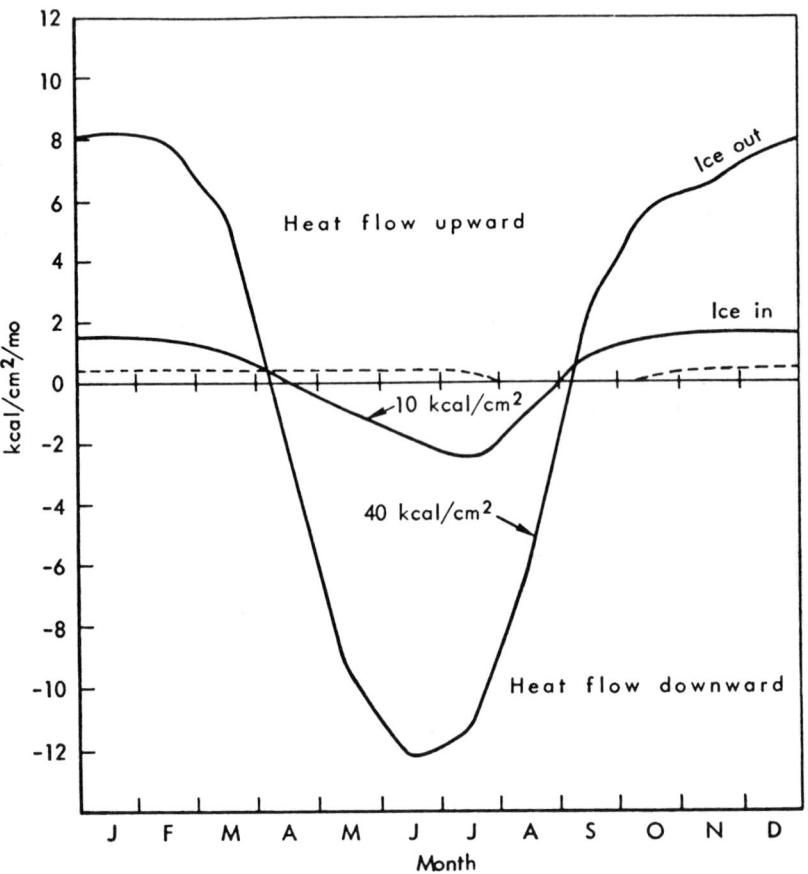

FIGURE 2 Heat flow to the surface from below under present ("ice-in") conditions and as estimated for an ice-free Arctic Ocean. (After Fletcher, 1965.)

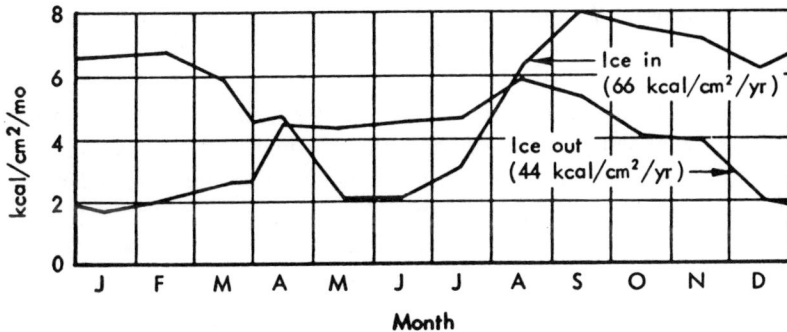

FIGURE 3 Atmospheric cooling in the central Arctic under "ice-in" and hypothetical "ice-out" conditions. (After Fletcher, 1965.)

about 10 kcal/cm^2, and this is enough to warm the ice mass and melt about a third of its thickness. A similar amount is supplied to the surface during winter, and most of this comes from cooling of the ice mass and some freezing. The actual upward flux of sensible heat from the liquid ocean is shown by the dotted curve.

The "ice-out" curve shows heat flow to the surface from below as estimated for an ice-free Arctic Ocean. According to these estimates, the ocean would gain about 40 kcal/cm^2 in summer and would lose a similar amount in winter, for a net balance of about zero. Of course, some rather uncertain assumptions must be made to obtain such estimates. Donn and Shaw (1965), for example, have estimated a somewhat smaller heat loss during winter, and, on this basis, they suggest that, if the pack ice were once removed, it would not soon return. Budyko (1962) came to a similar conclusion, but Badgley (1961) expresses the opposite view. I believe the question cannot be answered definitely until the uncertainties of estimation have been reduced and also until we have a better basis for judging how the atmospheric circulation would respond to such a condition. More about that later.

Now let us look at the annual variation of atmospheric heat loss over the Arctic basin under present conditions and as we estimate it would be if the Arctic Ocean were ice-free (Figure 3). The difference is enormous. Under present conditions, the atmosphere over the pack ice loses 6 to 8 kcal/cm^2/month during winter, but in summer it loses less than a third of this amount. Over an ice-free Arctic, the annual pattern would be almost the inverse, with strongest atmospheric cooling during summer and with winter values much smaller.

At this point we might ask, "How does the intensity of atmospheric heat loss in the polar region influence atmospheric circulation?" (Figure 4). Because the Arctic basin is centrally located with respect to the main planetary westerly circulation, we would expect that more intense atmospheric heat loss over the Arctic would mean stronger northward temperature gradients and stronger westerly winds around the periphery of the Arctic basin.

Figure 4 shows that such a relationship does indeed exist. In addition to our curve of the atmospheric heat loss, Figure 4 shows the mean index of zonal circulation at the periphery of the Arctic basin, computed for the years 1949 and 1962 by Putnins (1963). Allowing about 3 weeks for the atmosphere to cool and to respond to the thermal gradients, we see that the two curves correlate closely.

This brings us again to the questions, "What would the mean zonal circulation look like if the Arctic were ice free?" and "How would this affect atmospheric circulation at lower latitudes?"

To answer these questions, we must realistically model the entire planetary circulation under the assumption of an ice-free Arctic Ocean, but as yet this has not been adequately done. However, detailed calculations of zonal temperature distribution at various levels under conditions of an ice-free Arctic have been made by Rakipova (1966), using a theoretical model of zonal temperature distribution. According to these calculations, the intensity of both zonal and meridional circulation would decrease under these conditions, but more in winter than in summer, so that seasonal contrasts would be much smaller than with ice present. In high latitudes, poleward

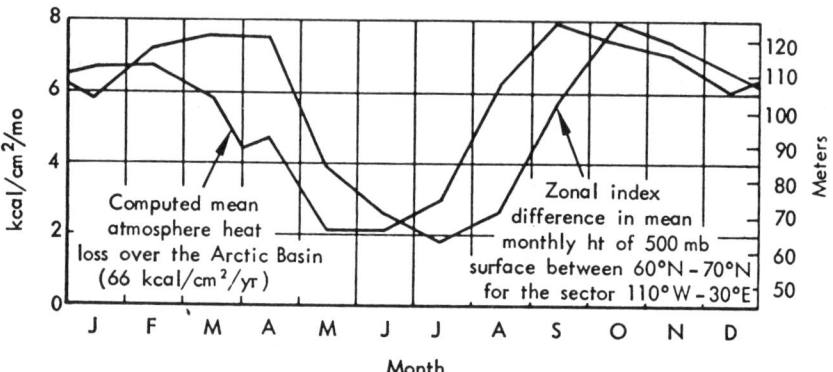

FIGURE 4 Atmospheric heat loss over the Arctic basin and zonal circulation intensity around its periphery. (After Putnins, 1963.)

atmospheric heat transport would decrease by about 25 percent during the cold half of the year. During the warm half of the year, poleward gradients would be a little greater in the upper half of the troposphere than in the lower half, implying stronger circulation, as we may deduce from Figure 3. Thus, these calculations suggest that an ice-free Arctic might indeed remain ice-free, but the uncertainties of the method are too great to draw firm conclusions.

NATURAL VARIATIONS IN THE EXTENT OF THE ARCTIC PACK ICE

We have seen that there are sound physical reasons why the extent of ice on the sea should influence atmospheric circulation, in addition to the traditional arguments about cyclones tending to follow the pack-ice boundary. Let us then consider briefly the following questions: "What makes the pack ice vary?" "How much does it vary from year to year?" "What long-term trends have been observed?"

Figure 5 illustrates some of the most important factors causing year-to-year variations. The upper curve shows the monthly values of solar radiation at the earth's surface, after taking into account reflection and depletion by the atmosphere above. The total annual value of solar radiation at the surface is about 73 kcal/cm^2, but the highly reflecting surface absorbs only about 18. Moreover, maximum absorption does not occur in June, when radiation intensity is greatest, but in July, when radiation intensity is rapidly waning. The reason for this is that collapse of the snow cover and puddling of meltwater on the ice causes a greater fraction of available radiation to be absorbed—but this occurs after the radiation intensity has already passed its maximum. If this event occurred 3 or 4 weeks earlier, it would make a large difference in the heat budget, and the dotted line is sketched to indicate the larger amount of heat that might be absorbed in such a case. Area, on this chart, represents a quantity of heat, and the square is drawn to show how much heat it would take to melt 2 m of ice, about two thirds of the total pack-ice thickness. It is clear that there is more than enough solar heat available at the surface to melt the whole ice pack if the absorptivity of the surface were sufficiently high.

Thus it can be seen that the total amount of melting in summer is extremely sensitive to factors that influence the date of first melting. A greater than usual number of cyclones advecting warm air during May and June is probably the most important factor in the melting of

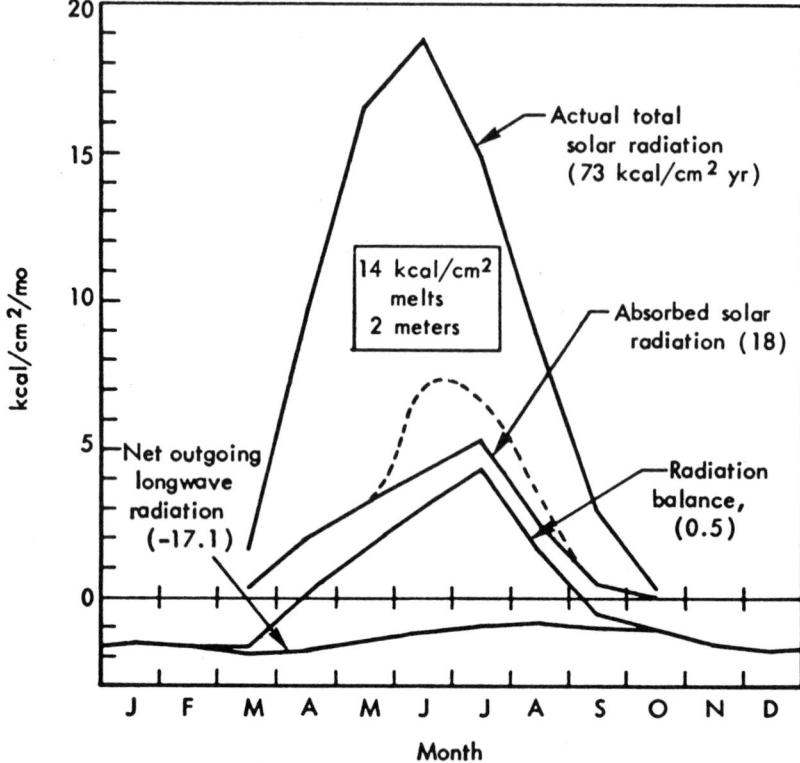

FIGURE 5 Radiation components at the surface in the central Arctic (long-term mean values).

the pack ice. A small positive air-temperature anomaly and higher wind speeds greatly increase turbulent heat flow from the air to the ice surface (Laikhtman, 1959). The resulting advance in melting date greatly increases absorption of solar heat.

If we examine the data from Arctic drifting stations over the last 15 years, we find that observed year-to-year variations are on the order of 8 kcal/cm^2, half the area of the square in Figure 5, and are primarily attributable to anomalous advection in early summer. We may therefore surmise that any general increase in the vigor of global circulation during May and June, whatever the planetary cause, would tend to cause melting of the Arctic pack ice.

On the other hand, anomalies of winter air temperature are much less effective in influencing ice thickness. Colder surface temperatures steepen the thermal gradient in the upper layers of snow, but freezing at the bottom of the ice is not drastically affected. To test the sensitivity of ice freezing and melting, Untersteiner (1964) constructed a theoretical model suitable for computer simulation of the thermal regime of pack ice under a wide range of boundary conditions.

Now that we have seen what makes the pack ice wax and wane, we may ask, "How much variation in thickness and extent have we actually observed?"

The answer is rather startling. According to Ahlmann (1945), between 1890 and 1940 the mean thickness of the ice decreased by about one third (from 365 cm to 218 cm), and in only 2 decades, from 1924 to 1944, the area covered decreased by about 400,000 mi^2.

How did the rest of the planetary system behave? During that half century (1890 to 1940) the intensity of the whole global circulation was increasing; in the North Atlantic sector of the circumpolar vortex, the westerly wind speed increased by some 20 percent, cyclones took more northerly paths, the Gulf Stream in the North Atlantic became stronger and passed farther north, and surface temperatures in the Norwegian Sea rose some 5.6°C (10°F). The whole planet seemed to undergo a slight warming, although warming was most pronounced in the North Atlantic and North Pacific, where cyclones were more frequent and more northerly in movement (Lamb, 1966).

As the speed of the westerly winds increased, the wavelength of the long waves in the westerlies also increased, displacing eastward the mean positions of troughs and ridges, with a corresponding displacement of rainfall patterns. Long summer droughts became more frequent in the south central parts of North America and Eurasia. In short, the changing pattern of global climate, including the warming Arctic, seemed to be the result of a general strengthening of global atmospheric circulation, the cause of which is still unknown.

Since 1940, much of this change has been reversed. The vigor of atmospheric circulation has been decreasing; ridges and troughs and rainfall patterns are farther west. The warm North Atlantic Current is weaker and farther south, and the edge of the pack ice has advanced considerably. Again, the cooling of the Arctic seems mostly to be due to a decreasing intensity of the circulation and a tendency for storm tracks to take more southerly paths especially in summer. Winter changes have been greatest in areas dominated by the Arctic anticyclones—namely, north central Canada and Siberia. North of Hudson

Bay, mean January temperatures have fallen more than 2.8°C (5°F) since 1940.

ATMOSPHERIC HEAT LOSS OVER ANTARCTIC REGIONS

Since these climatic changes are global in nature, let us look now to the Southern Hemisphere, and especially to Antarctica, for a more complete picture of how climate changes and for some clues about why climate changes. To do this, I have tried to study each component of the heat balance in order to construct a complete picture of the annual variation of all the important heat exchanges. In this way, we obtain a residual term in the heat balance that corresponds to the heat loss by the atmosphere. We also obtain a better understanding of just what physical factors are most influential in causing year-to-year variations, and we can examine the reasons for the natural variations of these factors and the possibilities for influencing them artificially.

We can divide the far south into three regions, in each of which the heat exchange processes are sufficiently uniform to permit us to describe the whole region by a single set of graphs. The three regions are the Antarctic continent, the ice-covered sea around the continent, and the ice-free sea around the continent. The graphs themselves are not discussed here, but the annual cooling patterns of the atmosphere obtained by this graphing method tell us an interesting story. Figure 6 synthesizes the patterns indicated by the graphs.

The time scale in Figure 6 has been displaced by a half year to make it easier to relate the cooling patterns to their counterparts in the Northern Hemisphere. Thus, Antarctic summer is in the center of the scale.

It can be seen that the annual pattern of heat loss by the atmosphere over the Antarctic continent, which is roughly the area of the Arctic basin, is very similar to that over the Antarctic pack ice, which at its maximum extent is about 1.5 times greater in area than the continent. The simple annual pattern is caused by (a) strong solar heating during summer, which partly compensates for the heat loss to space, so that atmospheric heat loss is reduced; (b) maximum atmospheric heat loss in winter, when there is no sun; and (c) rapid increase in heat loss in the fall, as solar energy wanes and loss to space is still high because of warmer temperatures. Over the continent, the atmospheric heat loss is considerably lower than over the pack ice. With so little heat reach-

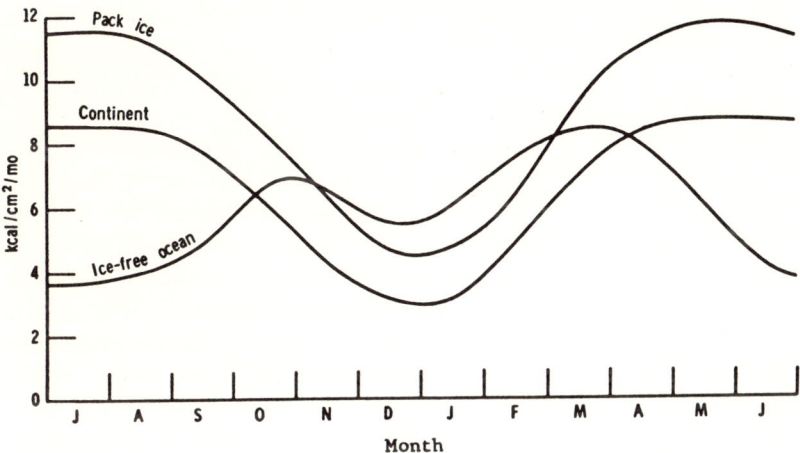

FIGURE 6   Annual pattern of heat loss by the atmosphere over the Antarctic continent, over the pack ice, and over the ice-free regions of the surrounding ocean.

ing the surface from below over the high, snow-covered Antarctic continent, the radiative loss from the surface simply lowers the temperatures of the surface until it attains approximate radiative equilibrium with the warmer atmosphere above. This means that very low surface temperatures, −30°C in the warmest month and −70°C in the coldest month, are typical. Thus, the heat loss to space is reduced by the generally lower temperatures of both surface and atmosphere.

Over ice-free ocean, the pattern of atmospheric heat loss is enormously different, both in shape and magnitude. In winter, most of the heat loss to space is supplied by the ocean, and atmospheric heat loss is small. During summer, this heat flux from the ocean is suppressed by the change in the air/surface temperature gradient, so that atmospheric heat loss increases, but the considerable absorption of solar heat by the atmosphere causes a second minimum in midsummer. The result is two maxima of atmospheric heat loss near the equinoxes. This general pattern, seen also in the estimate for an ice-free Arctic Ocean, is even more pronounced in the North Pacific and North Atlantic. The causal factors are high latitude, cold winter advection over ice-free ocean, and large solar absorption by a moist, cloudy atmosphere subjected to long hours of sunlight. I believe that when we are able to obtain better observational data for surface heat exchange and solar absorption in the atmosphere, it will become more apparent that this phenomenon, characteristic of only a narrow subpolar zone in both hemispheres, is an important cause of the long-observed but little-understood "semiannual oscillation" of the global atmosphere, which has often been attributed to some unknown cause in the equatorial region.

The important question is: "What difference does it make to the thermal forcing of atmospheric circulation if the Antarctic Ocean is ice-free rather than ice-covered?" As shown in Figure 6, the net heat loss of the atmosphere over an ice-free Antarctic Ocean in the winter is enormously smaller than the heat loss over continental regions or over an ice-covered ocean. In the summer, on the other hand, the net atmospheric heat loss over an ice-free ocean is somewhat greater than over continental or ice-covered oceanic regions. These differences in the patterns of atmospheric heat loss would, of course, have corresponding effects on the thermal forcing of atmospheric circulation.

Figure 7 illustrates that in the Antarctic, as in the Arctic, patterns of atmospheric heat loss are closely related to atmospheric circulation. In the upper part of Figure 7 we see that the mean winds over Mirny (Dolgina, 1967), at the periphery of the continent, closely follow the pattern of cooling over the interior.

It is difficult to find a suitable station from which to observe the characteristic effect of the ice-free polar ocean on atmospheric circulation, for most are either on the periphery of the continent or so far north as to be dominated by other processes. Argentine Island was chosen as most suitable, being far enough north to reflect maritime influences. The lower part of Figure 7 shows that, although at 500 mb the annual pattern of wind speed generally follows the "ice-covered" pattern, as at Mirny, wind speed at lower levels decreases during midwinter, presumably reflecting the influence of large heat flow from ice-free seas.

VARIATIONS IN THE EXTENT OF THE ANTARCTIC PACK ICE

The foregoing discussion attempts to explain why the extent of ice on the sea is a sensitive "climatic level" that regulates heat exchange between the ocean and atmosphere in both the Arctic and the Antarctic. Let us now attempt to relate the magnitude of these effects to the entire planetary ocean–atmosphere system. In the Arctic, the annual variation in the area covered by pack ice is about 20–25 percent, and the long-term variation in maximum ice cover during the last century has been about 10–15 percent. In the Antarctic, the maximum area of

Ice on the Ocean and World Climate   15

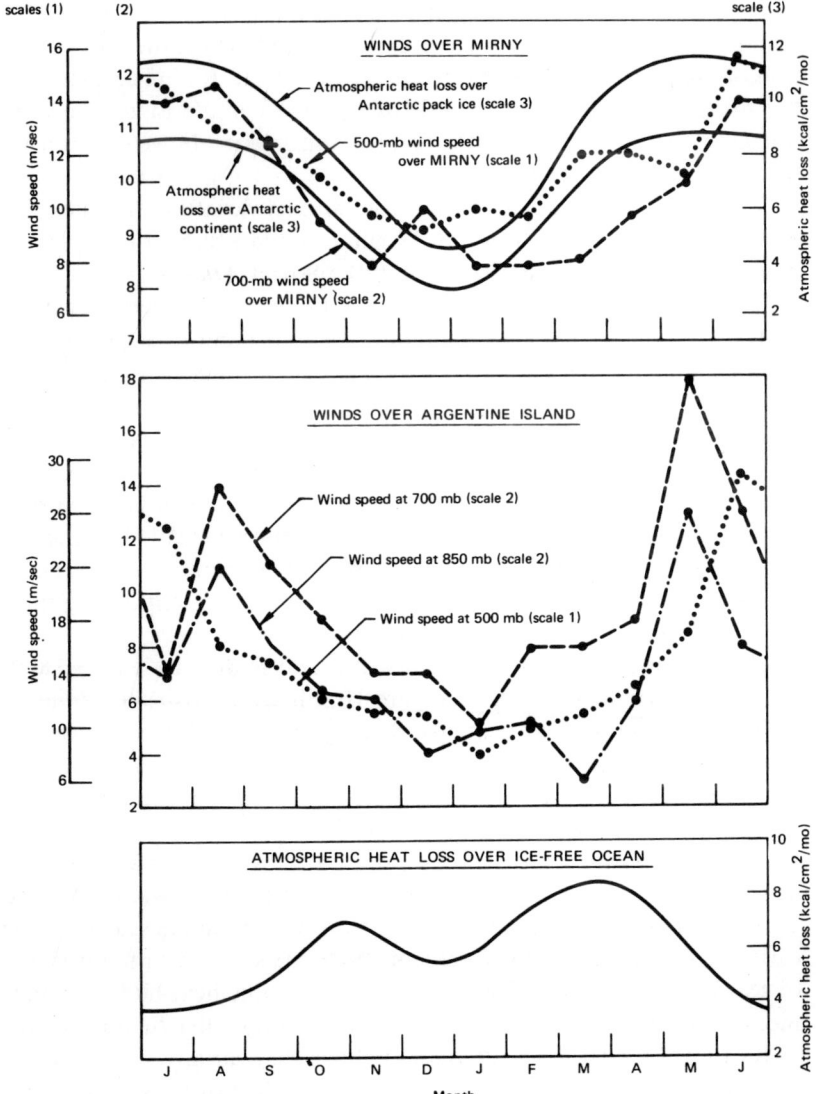

FIGURE 7 Wind speeds over Mirny and over Argentine Island and their relationship to atmospheric heat loss over the continent, pack ice, and open water.

pack ice is more than 1.5 times that in the Arctic, and the annual variation of the area of pack ice in the Antarctic is 85 percent of the maximum pack-ice area (according to Treshnikov, 1967, from 18.8×10^6 km^2 in September to 2.6×10^6 km^2 in March). Thus, the annual variation of area covered is some 6 times greater in the Antarctic than in the Arctic. The relative magnitude of year-to-year and long-term variations may be correspondingly large, but unfortunately we have very few data on these variations.

What is the pattern of this enormous annual variation of the extent of ice on the sea in the Antarctic? Figures 8 and 9 show the mean monthly pack-ice boundaries given in the Soviet *Atlas of Antarctica* (1966). They tell an interesting story, but one should keep in mind that, because large year-to-year variations do occur and because observational data are meager, such maps cannot be viewed with much confidence.

Of most significance is that the greatest change in ice extent occurs during December, when solar radiation is at a maximum. It is obvious from Figure 9 that, since the presence of pack ice in summer somewhat increases atmospheric heating and greatly reduces oceanic heating, large anomalies of summer heating of the Antarctic Ocean can occur, and such anomalies could be expected to influence ice extent during the following winter.

The extreme sensitivity of the oceanic heat budget to seasonal anomalies of ice extent is shown by Figure 10. Figure 10 also shows the annual variation of solar heat absorbed by the area of ocean corresponding to the maximum ice extent (20×10^{16} cm^2) under the assumption that pack ice absorbs 45 percent of incident radiation and that ice-free ocean absorbs 90 percent. To show the effect of year-to-year variations in ice extent, the computation was repeated with the pack-ice curve advanced by 2 weeks and retarded by 2 weeks. A season 2 weeks late reduced oceanic heating by 4 percent and a season 2 weeks early increased heating by 7 percent. It can be seen that the influence of seasonal anomalies of ice extent on the oceanic heat budget is enormous, especially when the ice boundary shrinks earlier than usual.

It can be seen from the discussion above that variations in ice extent on the southern ocean can amplify the effect of long-term changes in solar radiation intensity. An increase in the solar constant tends to hasten the retreat of the ice, causing a larger fraction of the higher-intensity radiation to be absorbed by the ocean. Conversely, a decrease in solar intensity delays the retreat of the ice, reducing the fraction of

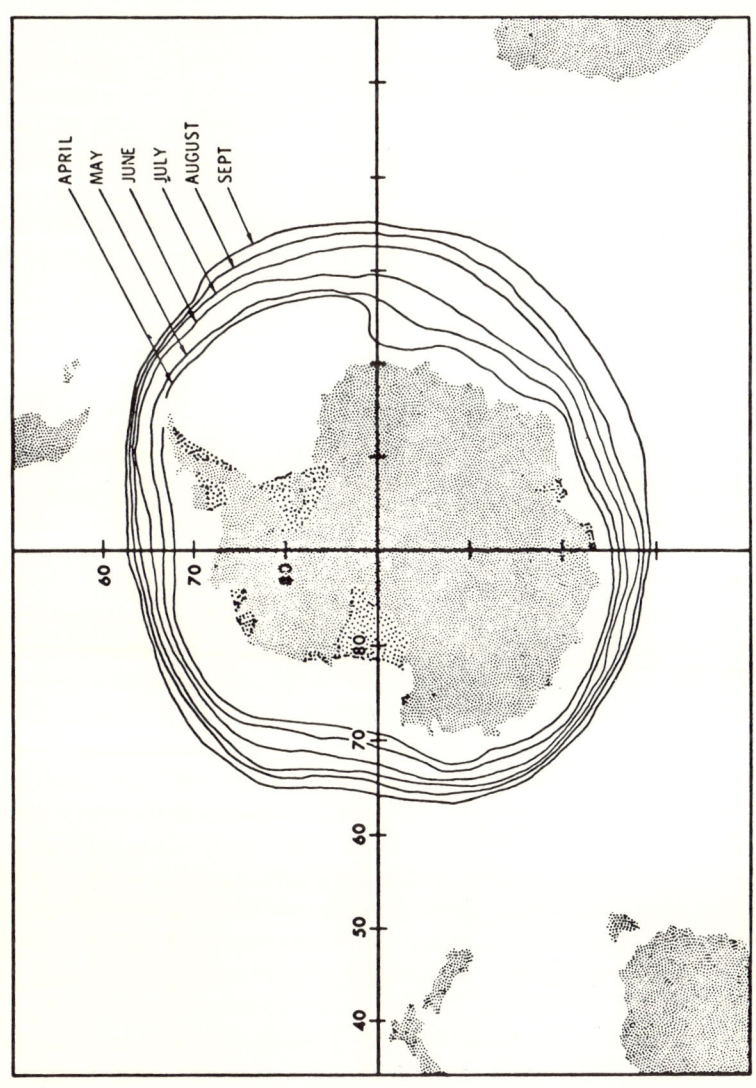

FIGURE 8 Monthly variations in the boundary of the pack ice in the Antarctic winter. (Adapted from "Atlas of Antarctica," 1960.)

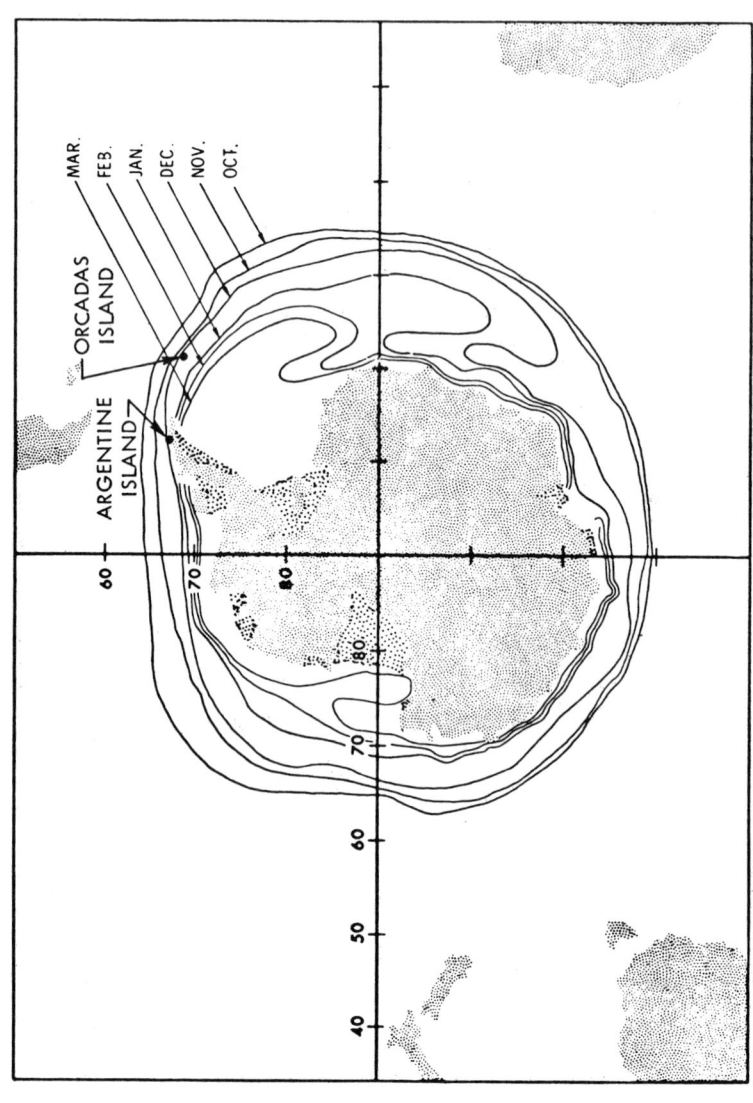

FIGURE 9 Monthly variations in the boundary of the pack ice in the Antarctic summer. (Adapted from "Atlas of Antarctica," 1960.)

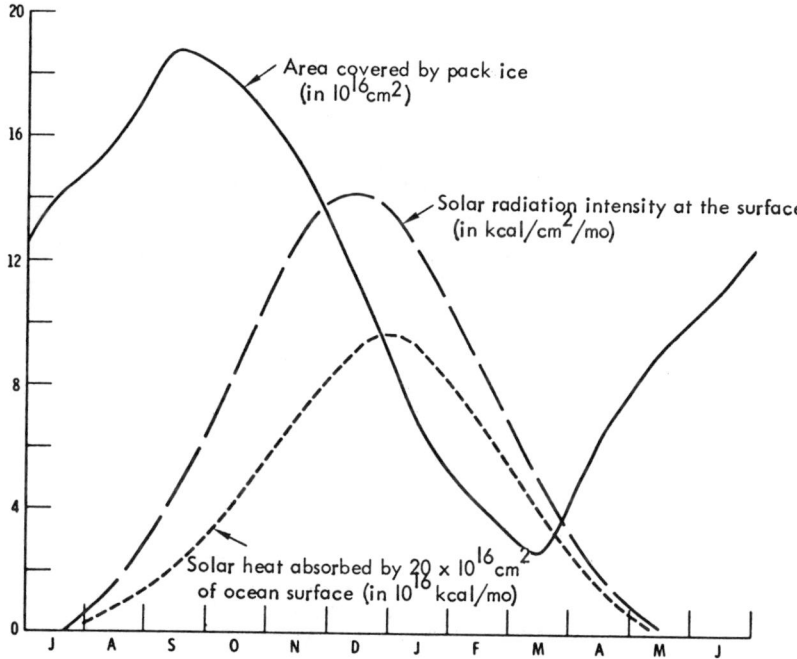

FIGURE 10 Annual variation of pack-ice area and solar heat absorbed at the surface in the Antarctic. (Monthly pack-ice values after Treshnikov, 1967.)

the weaker solar radiation absorbed and thus greatly reducing the total heat absorbed.

In winter, on the other hand, the ice cover affects the oceanic heat budget in the opposite sense; it suppresses heat loss by the ocean—i.e., the more ice cover, the less heat loss. Quantitative estimates for winter are more uncertain, however, because the extent and location of open leads and polynyas can drastically alter the picture.

ANTARCTIC COOLING AND GLOBAL CIRCULATION

We have seen that a major effect of ice on the sea is to greatly increase atmospheric cooling during winter. We may now ask, "How might variations in sea-ice extent influence global climate?" (see Figure 1). The motion of the atmosphere is a direct result of the global pattern of heating and cooling, which produces potential energy; this potential energy is, in turn, transformed into kinetic energy by atmospheric motion, maintaining the general circulation against frictional dissipation and transporting heat from regions of heat surplus to regions of heat deficit.

These are very complex processes, but we can illustrate some of the gross features of global circulation and how it is influenced by Antarctica by means of Figure 11, which is based on data gathered during the International Geophysical Year (IGY) (adapted from Borisenkov, 1965). The upper curves show the latitudinal distribution of potential energy from pole to pole. In the equatorial regions, the atmosphere

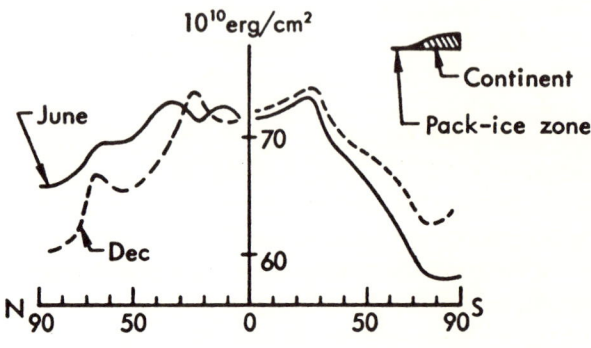

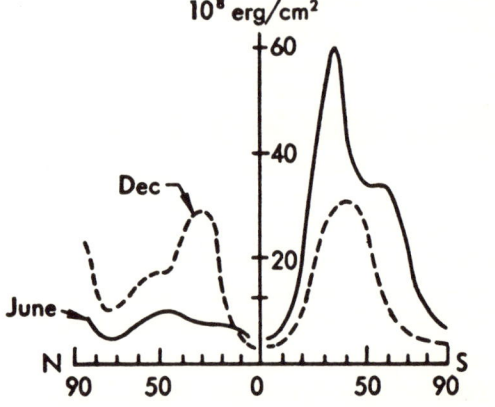

FIGURE 11 Global distribution of potential energy (upper curves) and kinetic energy (lower curves) in summer and winter. (Adapted from Borisenkov, 1965.)

is warm, its center of gravity is high, and potential energy is at a maximum. The polar regions are the main heat sinks of the global system, and the center of gravity of the colder polar atmosphere is lower than that of the equatorial atmosphere. In December, the potential energy is about equal in the two hemispheres, and poleward gradients are also very similar. Temperatures in both polar regions are about −30°C. The lower part of Figure 11 shows that in December kinetic energy is also about the same for the two hemispheres.

During June the picture is dramatically different. Potential energy is slightly greater in the Northern Hemisphere, but, with relatively warm Arctic summer temperatures, the poleward gradients are very much smaller than those in the Southern Hemisphere, where the atmosphere is losing heat at a high rate over both the Antarctic continent and over the pack ice. As a result, the kinetic energy of the Southern Hemisphere circulation in June is about 3.5 times greater than that of the Northern Hemisphere. In the fall, the meteorological equator is at its maximum distance north of the real equator. Just how much the Southern Hemisphere circulation feeds energy to the Northern Hemisphere circulation, either by transport of latent heat or by air-mass exchange, is not well understood, but we do know that evaporation is proportional to wind speed and that moisture transport from the Southern Hemisphere is an important energy source for the Northern Hemisphere circulation. Thus, it is tempting to surmise that the general strength of the whole global circulation will be influenced by variations in Southern Hemisphere circulation, and that the extent of ice on the southern ocean, in turn, is a significant influence on these variations in Southern Hemisphere circulation.

Variations in general circulation during the last two centuries, as comprehensively presented by Lamb (1966), reveal that during a period of increasingly vigorous global atmospheric circulation, the two hemispheres did not behave symmetrically, with belts of high and low pressure displaced correspondingly toward or away from the equator. Instead, we find a displacement in the same direction in both hemispheres, suggesting that the Southern Hemisphere circulation is the controlling factor. Moreover, when the extent of Antarctic ice is greatest, global circulation seems to be strongest. Thus, in the Southern Hemisphere, these two factors are acting in the same direction, whereas in the Arctic, the relationship is reversed. The Arctic was much colder in 1800 than in 1930, atmospheric circulation was weaker in both hemispheres, and Arctic ice was more extensive. Antarctic pack ice, however, seems to have been less extensive in 1800 and the southern low-pressure belt somewhat farther south. According to Lamb (1966), the latitudinal shift of the Southern Hemisphere high-pressure zone has been about 3° to 5°. Since such latitudinal shifts are quite resistant to changes in thermal forcing (Sawyer, 1966), an even greater latitudinal change in ice extent is implied. At 60°S, the approximate ice border, a zone 1° of latitude in thickness corresponds to about 1 percent of the entire area of the Southern Hemisphere. Such a drastic change as the absence or presence of ice over perhaps 5 percent of the hemisphere suggests a large change in thermal forcing, both in annual values and seasonal pattern.

Thus, the hypothesis that emerges is one in which changes in solar-radiation intensity in Antarctica are amplified by variations in the extent of sea ice, causing variations in the Southern Hemisphere circulation, and with the weaker Northern Hemisphere circulation following the trends of the Southern Hemisphere, especially during the northern summer. The extent of sea ice in the north is also an influential lever on local atmospheric behavior, but its waxing and waning, which, as noted earlier, is most sensitive to warm-air advection in summer, is basically a response to the vigor of global circulation during May, June, and July, the Southern Hemisphere's winter. The waxing and waning of the sea ice in the Northern and Southern Hemispheres occurs asynchronously. In both polar regions, more extensive sea ice tends to intensify atmospheric cooling and increase circulation. In the Southern Hemisphere, these factors seem to act in the same direction, but in the Northern Hemisphere the ice extent is more influenced by, rather than influencing, the global system.

In this connection, the deep cores collected on the Antarctic continent may provide one of our best long-term records of global circulation intensity. More intense circulation seems to bring more intense cyclones into the interior, causing more rapid snow deposition there. Much more work will be needed if we are to fully understand these interactions, but their significance is obvious (Barkov and Petrov, 1966).

SECULAR TRENDS OF ANTARCTIC SEA-ICE EXTENT

Unfortunately, little is known about year-to-year variations in the extent of Antarctic sea ice, even for the years since the IGY (Heap, 1964). There is only one long-time series of systematic and homogeneous data from which to draw inferences about the variations of ice cover of antarctic seas. This data comes from the Orcadas station, on Laurie

Island in the South Orkneys (see Figure 9). The station, established by Great Britain in 1903 and later taken over by Argentina, has provided almost uninterrupted observations of meteorological and ice conditions since 1903 (Schwerdtfeger, 1959). Orcadas station is well situated to observe the general ice conditions of the southern ocean, as it is about midway between the maximum and minimum ice boundaries.

Since 1903, systematic observations have been made of the dates on which Scotia Bay (on the south side of the island, facing the Weddell Sea) was closed by ice in the fall and opened again the following summer. These data are presented in Figure 12. It can be seen that the iciness has varied widely. In several years, the Bay was not closed by ice until late July, and in 1965 not until August 5. In other years the Bay was closed in April, and throughout one summer (1928–1929) the Bay did not open at all.

Variations in the dates of ice breakup in summer are much larger than variations in the dates of freeze-up in the fall. This may be due in part to the erratic occurrence of wind conditions favorable to breakup. On the other hand, if the general trends reflect large-scale variations in ice extent, the corresponding variations in heat budgets for the ocean and atmosphere are large, as shown in Figures 10 and 13.

In general, the trend seems to have been toward more iciness up to about 1928, and less iciness since then. The period from about 1920 to 1933 was particularly icy, as was to a lesser extent the 1912–1916 period. The 1930's and 1940's were less icy, and the 1950's and 1960's much less icy.

These periods generally correspond to climatic epochs whose manifestations in the Northern Hemisphere have been studied in detail by various authors. The present study does not attempt to explore the basic physical causes of these planetwide climatic variations; rather, it attempts to clarify the interrelation among heat budget, ice extent, and atmospheric circulation for the high southern latitudes. For discussions of possible physical causes of planetary variations, the reader is referred to Willett (1951, 1953, 1961, 1964, 1965), who attributes planetary climatic variations to variations in the composition and intensity of solar radiation, or to Budyko (1961, 1964, 1967, 1968), who attributes these climatic variations primarily to variations in atmospheric transparency caused by volcanic eruptions and industrial pollution. To facilitate comparison with the works of these authors, Figure 12 also includes the times of major and minor sunspot maxima given by Willett and the times of major volcanic eruptions mentioned by Budyko.

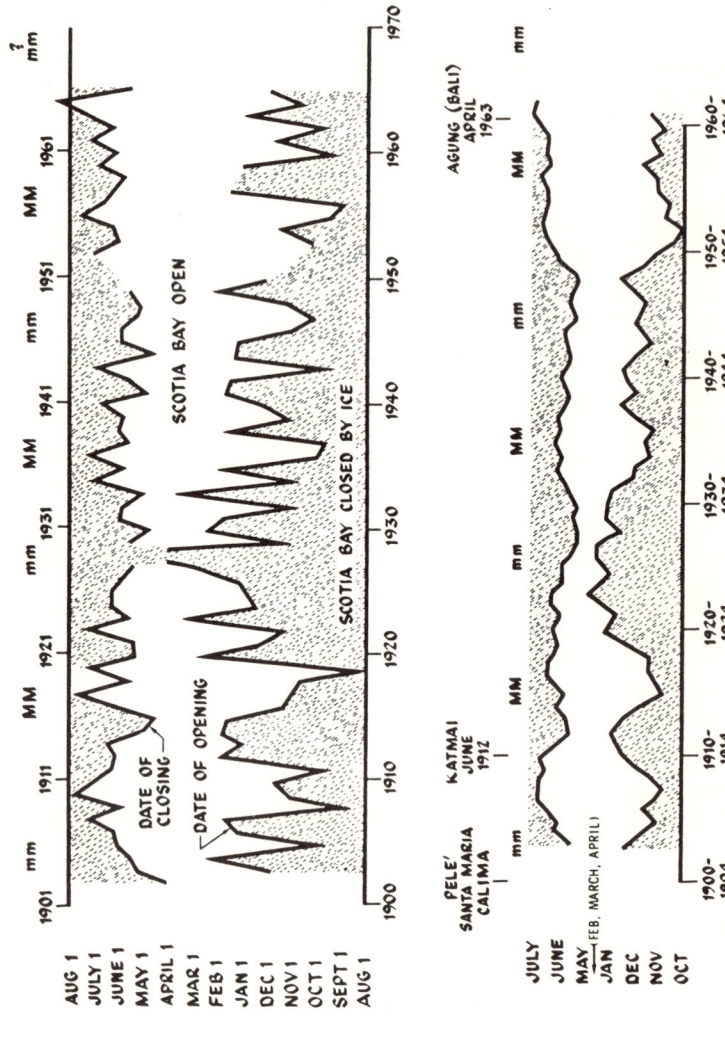

FIGURE 12 Trends of iciness in the Weddell Sea, as indicated by dates on which Scotia Bay on Orcadas Island (Figure 9) was closed by ice in fall and open in summer. Upper curves are unsmoothed data, lower curves are 5-year moving averages. Major volcanic eruptions are indicated on the lower graph; MM and mm are times of major and minor sunspot maxima, respectively. (Data from Schwerdtfeger, 1959.)

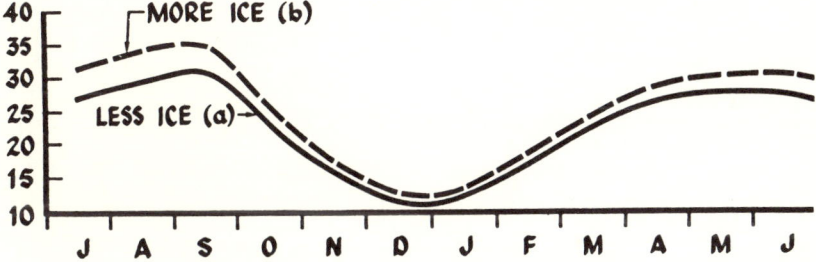

FIGURE 13   Annual variation of atmospheric heat loss south of 60°S; under the assumption that although maximum ice extent is unchanged, the date of formation and dissipation is displaced one month toward less iciness (a) or one month toward more iciness (b).

Interpretation of the iciness data in Figure 12 is facilitated by reference to Figure 14, which presents secular trends of mean monthly air temperature at Orcadas for each month of the year. Figure 14, in turn, can be interpreted better by referring also to Figures 8 and 9 for the months of most rapid change in ice extent. From these figures, it may be noted that

Secular variations of surface air temperature for the interval December through February are relatively small because of the stabilizing effect of melting ice and cold oceans during the warm season.

For the transitional months of April–May and September–October, the cooling trend culminating in the 1920's is clearly expressed. The parallel changes of iciness can be noted in Figure 12.

The secular changes are greatest in the coldest months (June through August). During the period of greatest iciness in the late 1920's, winter minimum temperatures occurred in June (month of the winter solstice), but a decade later the warmer winter minimums occurred in August. This feature suggests variation in solar radiation as a basic causal factor.

June temperatures have risen sharply since about 1928, with a small regression in the early 1940's. This regression, as well as the similar minimum about 1915, corresponds to icy Junes, as shown in Figure 12.

August temperatures have risen sharply since the mid-1930's and do not reflect the regression in the 1940's.

Thus Figures 12 and 14 present a coherent picture of secular trends at Orcadas. To what extent these trends are representative of the southern ocean in general is still unclear.

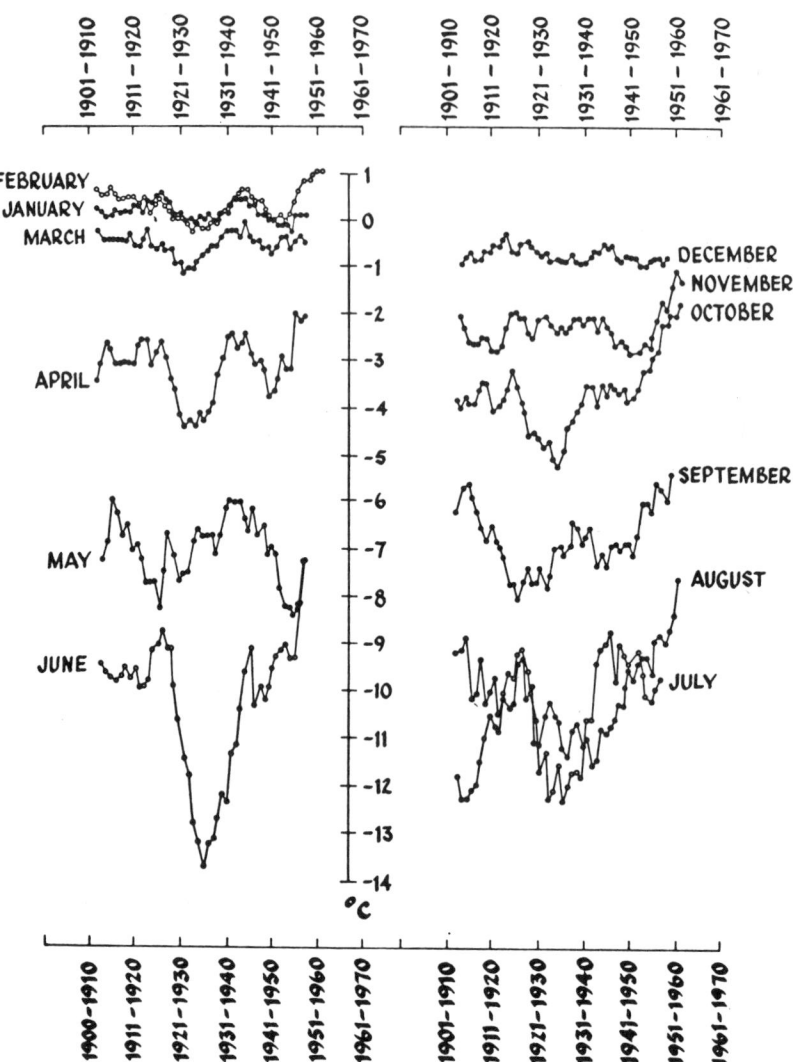

FIGURE 14 Trends of mean monthly surface air temperature at Orcadas as shown in 10-year moving averages. (Adapted from Rubinstein and Polozova, 1966.)

SECULAR TRENDS OF ATMOSPHERIC CIRCULATION INTENSITY OVER THE SOUTHERN OCEAN

If secular variations of Antarctic ice extent are associated with secular variations of the intensity of atmospheric circulation, and *if* the data for Orcadas are indicative of the general iciness of the southern ocean, then it should be possible to find some correlation between the Orcadas ice data and appropriate indices of the intensity of zonal circulation.

Unfortunately, it is difficult to find long-time series of appropriate indices of the intensity of general atmospheric circulation for the Southern Hemisphere. The two potential sources of data on north–south pressure gradients are South America and New Zealand. The longest time series, and probably the most reliable data, are from New Zealand and are presented for January and July of each year by Lamb and Johnson (1966). From Figure 12, we see that closing of the Bay occurs during May–June–July, so the secular curve of closing dates should be representative of winter severity and should show some correlation with trends of circulation intensity for July if, indeed, circulation intensity is related to iciness. Figure 15 shows the circulation indices for July, in the antarctic midwinter, superimposed on the curve of freeze-up dates at Orcadas. The curves seem to correspond well with each other for the period of record, with circulation intensity increasing during periods of increasing iciness and decreasing during periods of decreasing iciness. A correlation coefficient of .17 for the unsmoothed yearly curves indicates a 10 percent probability of a chance relationship.

In short, there does seem to be a significant correlation between Southern Hemisphere zonal circulation intensity in winter and iciness of the southern ocean, but the questionable representativeness of both indices used calls for further verification.

Figure 16 shows a similar comparison of summer ice conditions and summer (January) circulation intensity. The two curves show no significant correlation (correlation coefficient .07). This result is consistent with the reasoning described earlier in the discussion of Figure 13. That is to say, in winter the presence or absence of ice makes a large difference to the atmospheric heat budget, because an ice cover cuts off the large heat flux from ocean to atmosphere, but in summer the atmospheric heat budget is not much affected by ice extent. In summer, the presence or absence of ice makes a big difference to the heat budget of the *ocean,* because an ice cover reflects much of the solar energy. But heat flux from the ocean to the atmosphere during summer is so small that the presence of ice makes little difference to the *atmospheric* heat

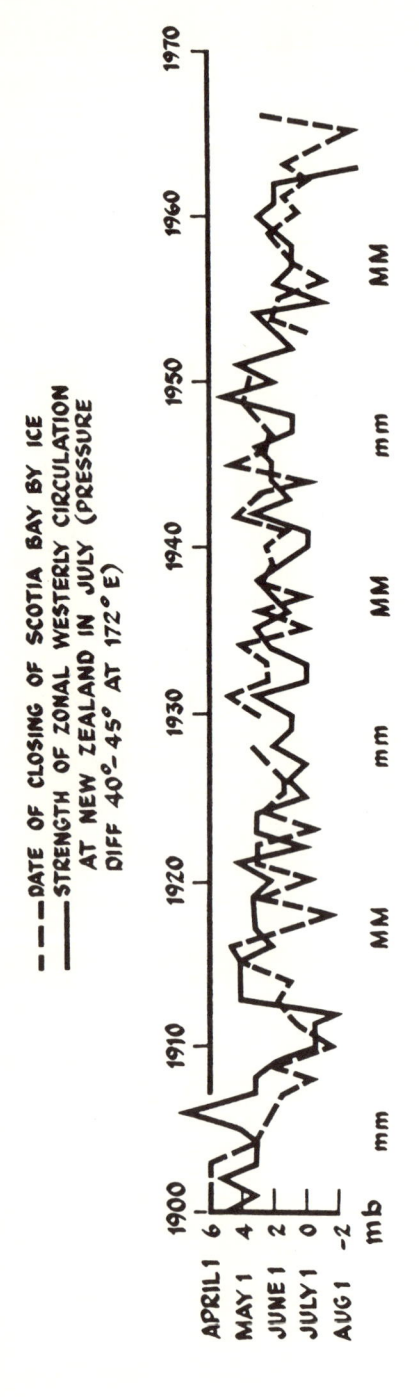

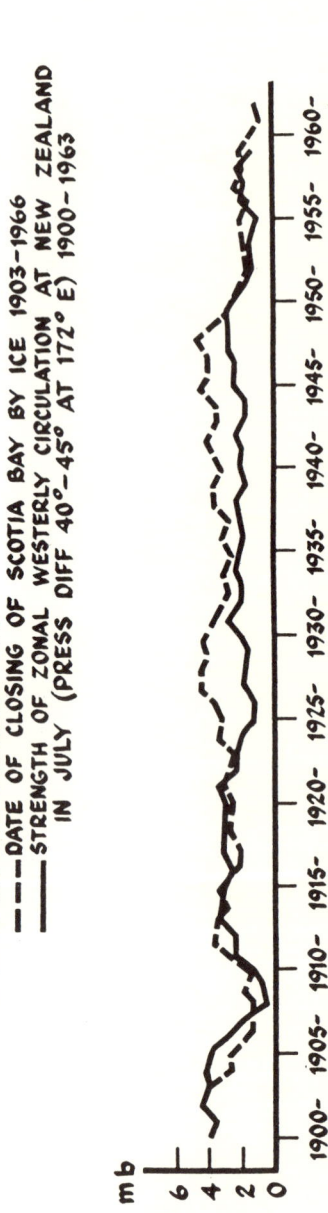

FIGURE 15 Yearly indices of Antarctic iciness and zonal circulation intensity in July at New Zealand. (Iciness data from Figure 12; pressure data from Lamb and Johnson, 1966.)

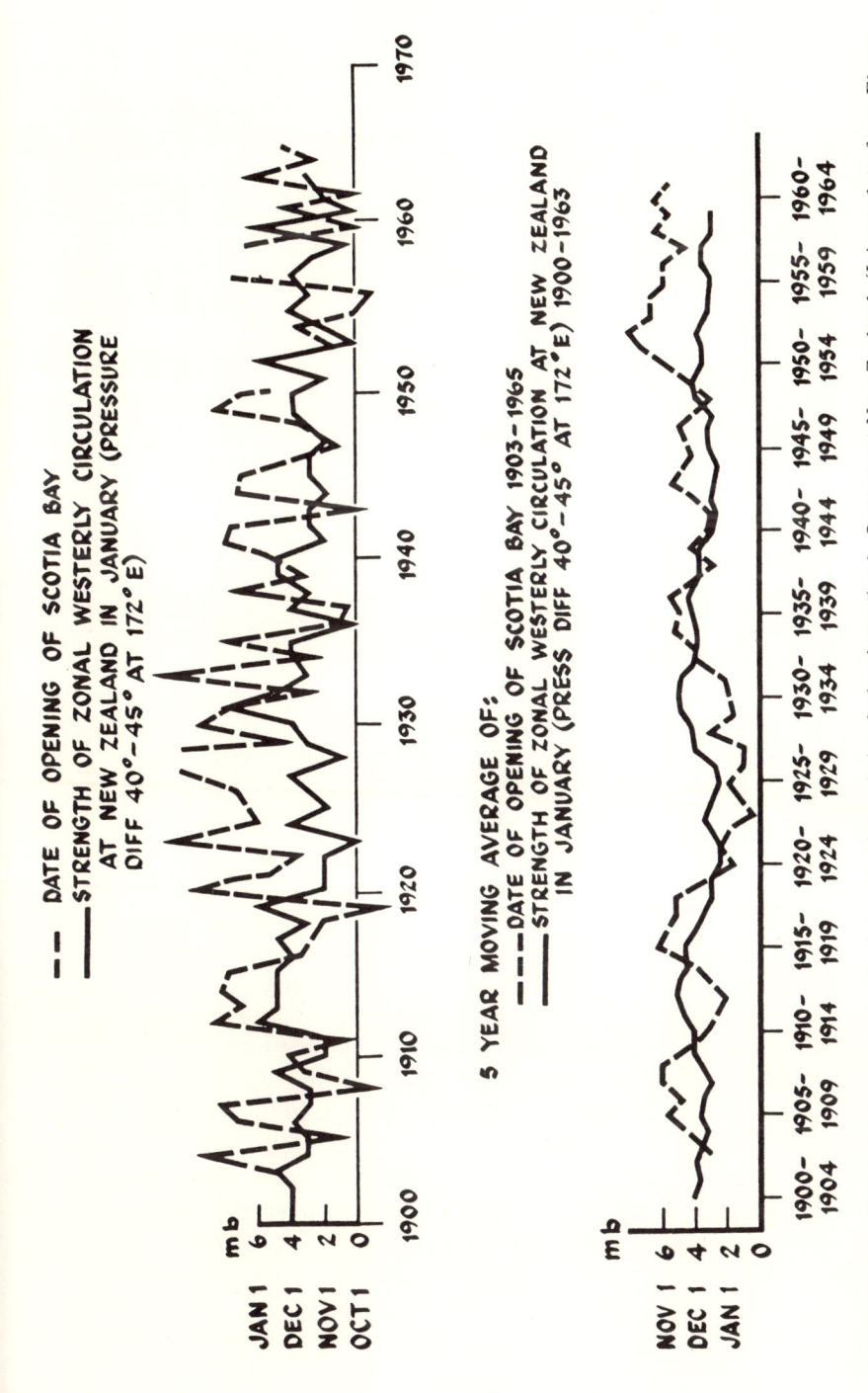

FIGURE 16 Yearly indices of Antarctic iciness and zonal circulation intensity in January at New Zealand. (Iciness data from Figure 12; pressure data from Lamb and Johnson, 1966.)

budget—slightly reducing the heat flow from the ocean and slightly increasing solar absorption in the atmosphere. It is therefore to be expected that variations in summer ice cover would have little immediate influence on thermal forcing and zonal circulation intensity. Alteration of the oceanic heat budget would presumably be felt by the atmosphere later, during the subsequent fall (April).

CONCURRENT CHANGES IN GLOBAL CLIMATE

For detailed descriptions of climatic changes in various regions of the earth during the present century, the reader may refer to Mitchell (1963), Lamb (1966), or Rubinstein and Polozova (1966). What emerges is, in summary, worldwide warming from the 1880's or earlier, through the 1930's, followed by worldwide cooling since about 1940, but more rapid in the 1950's and 1960's. The warming was most pronounced in northern oceanic areas, especially the North Atlantic, and the cooling was most pronounced in continental areas of northern Canada and Siberia. The general trends for the Northern Hemisphere are indicated by curves 1 and 2 of Figure 17. These curves were calculated from monthly maps of temperature anomalies over the Northern Hemisphere from 1881 to 1960 (Budyko, 1968). They can be taken as representative of the Northern Hemisphere as a whole, but not necessarily for all individual regions. For example, the North Atlantic region continued to warm until the 1950's before the trend reversed.

It is apparent from the preceding discussions that warming and cooling of the climate in the Orcadas region occurred approximately in the sense opposite to that of the global trends. Just how well Orcadas trends represent those of the remainder of the Antarctic region needs further investigation, but present evidence suggests that Orcadas trends do correspond to the trends of general Antarctic ice extent (Barkov and Petrov, 1966; Petrov, 1964; Kotlyakov, 1962). That Orcadas data reflect climatic variations over all of Antarctica is indicated in Figure 17 by the remarkable similarity of trends of snow accumulation at the South Pole to trends of Orcadas iciness. Both are presumably related to variations in atmospheric circulation intensity, with more intense circulation and a greater number of deep cyclones penetrating to the pole area when Antarctic sea ice was more extensive. Thus, the so-called "worldwide" warming and cooling trends were not really worldwide, but were more like an oscillation in which the high southern latitudes behaved in a sense opposite to that of the rest of the planet, and especially to that of the high northern Arctic regions. Figure 18 shows

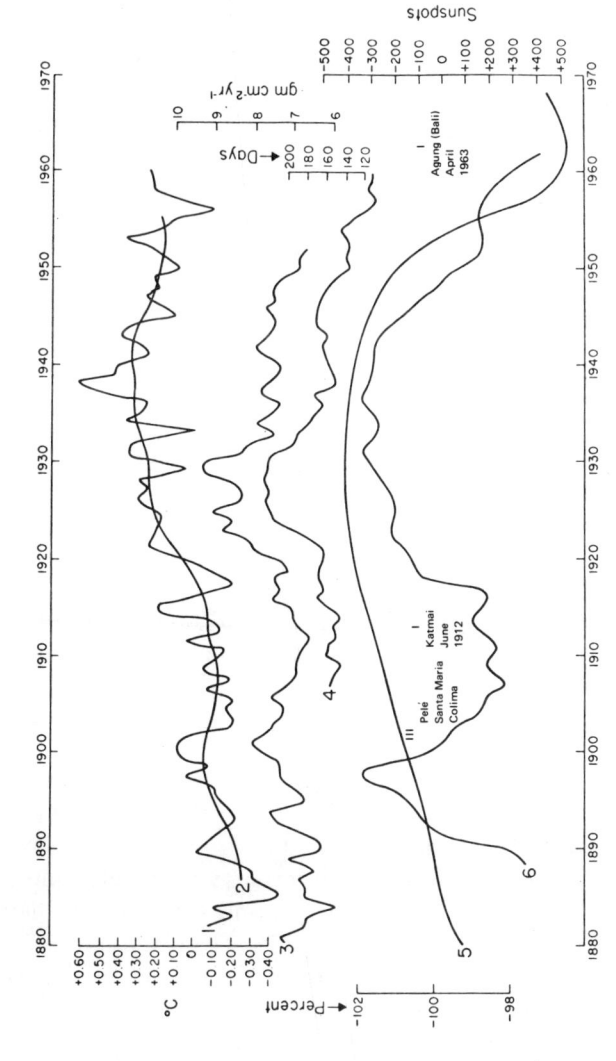

FIGURE 17 Yearly indices of iciness of Antarctic and Arctic Seas, of snow accumulation at the South Pole, and of Northern Hemisphere temperature. (1) Unsmoothed and (2) smoothed 10-year moving average of deviation from mean annual temperature in the Northern Hemisphere, from monthly maps of anomalies, 1881–1960 (Budyko, 1968); (3) 10-year moving average of annual snow accumulation at the South Pole (Giovinetto and Schwerdtfeger, 1966); (4) iciness of the Weddell Sea, Antarctica, 10-year moving average of the number of days with ice on the bay at Orcadas (Schwerdtfeger, 1959); (5) cumulative deviations from mean number of sunspots (Nazarov, 1963); (6) Variation of direct solar radiation with cloudless sky from stations in Europe and America (Budyko, 1968).

this relationship in greater detail. The upper curves show variations in frequency of zonal as opposed to meridional types of atmospheric circulation, and the lower curves show iciness of the Weddell Sea. Various authors have shown that observed climatic trends in the Northern Hemisphere are most closely related to the frequency of zonal circulation. Figure 18 shows that, since 1900, trends toward greater zonality of the Northern Hemisphere circulation have generally coincided with trends toward greater iciness of the Weddell Sea, as have the reversals of these trends—with variations in Northern Hemisphere circulation lagging behind variations in the southern ocean by about 5 years. It is thus of particular interest also to compare longer-term Antarctic trends and individual anomalous years with trends in high northern latitudes and to identify some features of global atmospheric circulation associated with these changes.

Long-Term Variations

With regard to longer-term climatic variations it would be especially pertinent to know how the antarctic climate varied during the "Little Ice Age," which culminated in the Northern Hemisphere at about AD 1700 (Lamb, 1966). At that time, according to Lamb, zonal flow at midnorthern latitudes was weaker and farther south than it is now by 3° to 5°.

There are, of course, no meteorological records of this period for the far south, but the close correlation, shown in Figure 17, among snow accumulation at the South Pole, iciness in the Weddell Sea, and climatic trends in the Northern Hemisphere suggests that the rate of snow accumulation at the pole may be a good index of variations in global atmospheric circulation. Snow cores can be dated to the Little Ice Age and to the warm period from AD 1100 to 1300, as well. As yet, however, the data are fragmentary.

In Figure 19, the graph of 22-year means given by Giovinetto and Schwerdtfeger (1966) is extended backward in time by including their more uncertain short-period samples. The resulting general impression is of a minimum accumulation during the Little Ice Age. If the relationships shown in Figure 18 existed at that time, then the Little Ice Age was a warm period at Orcadas and perhaps for the southern ocean.

Short-Term Variations and Individual Anomalous Years

From Figure 12, it may be seen that the most extreme ice years at Orcadas were 1928, when Scotia Bay failed to open at all, and 1965,

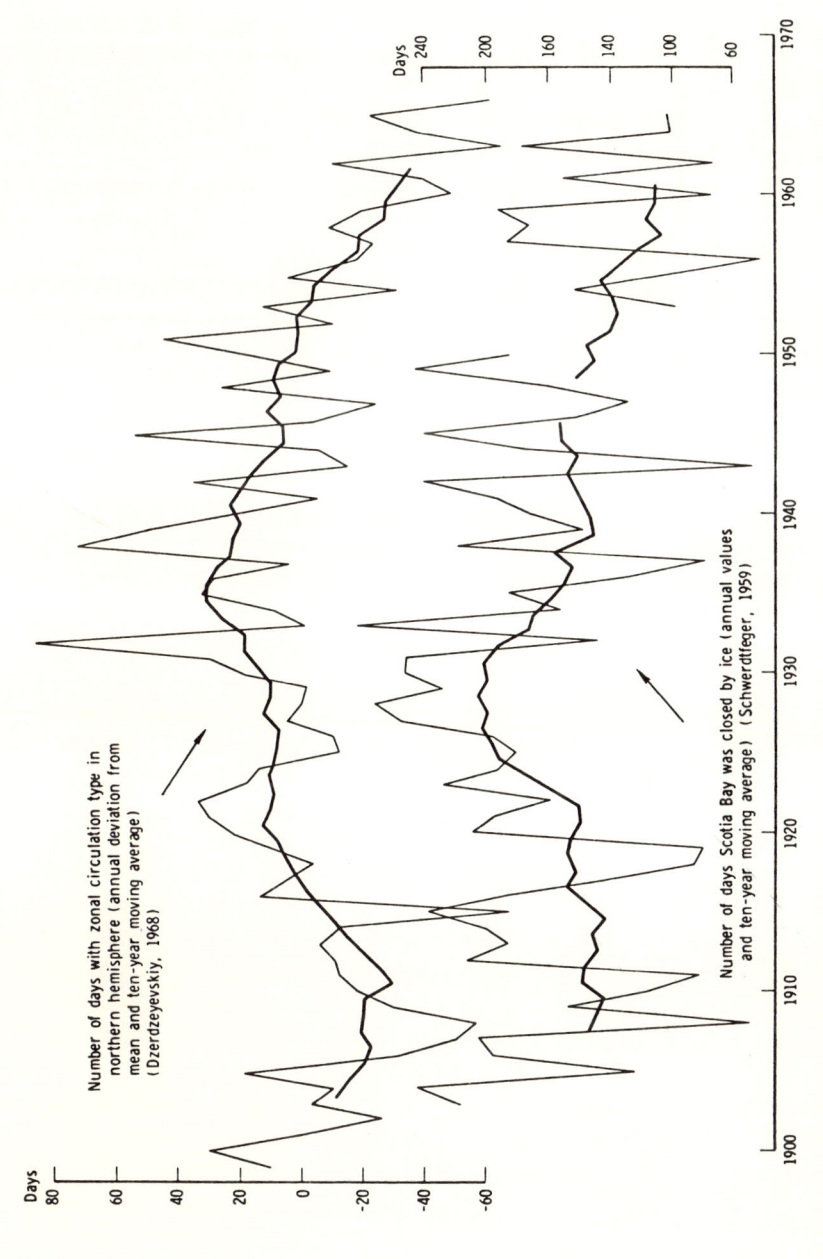

FIGURE 18 Variations in the basic character of Northern Hemisphere atmospheric circulation and in the iciness of the Weddell Sea.

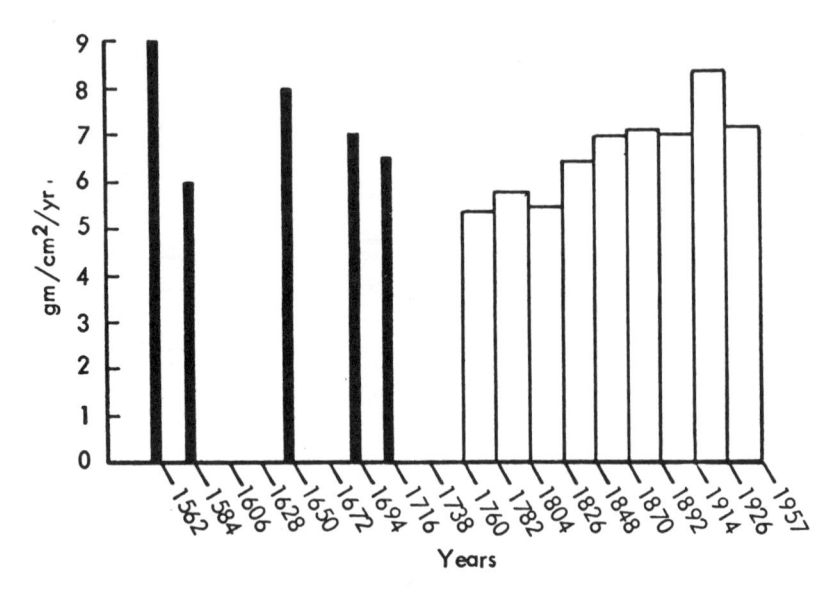

34 BENEFICIAL MODIFICATIONS OF THE MARINE ENVIRONMENT

FIGURE 19 Snow accumulation at the South Pole during the "Little Ice Age" (after 1760, 22-year means; before 1760, short-period samples). (Adapted from Giovinetto and Schwerdtfeger, 1966.)

when it was not closed by ice until August 5. It is of interest to examine climatic behavior in the Arctic at these times. The general relationship is shown by Figure 20, in which late winter temperatures at Orcadas are compared with winter temperatures at Archangel, on the White Sea. The Archangel temperature curve and the following description of meteorological conditions were presented by Yakovlev (1966):

The winter of 1965-1966, especially its second half, was noted for its persistent and very intense frosts. The mean temperature for all the winter months was substantially lower than the norm. . . .

Such an anomalously cold winter had never once been observed during a period from 1878 through 1965. The coldest winter noted had been in 1892-93. Its mean temperature was $-14.2°$. The mean temperature of the winter of 1965-66 equalled $-15.5°$. . . . [T]he coldest months were February and March, with their mean temperature dropping more than $10°$ below the norm. Such cold months had never been observed here during the period from 1878 through 1965. . . .

The very low temperature of the whole winter of 1965-66, especially its second half, was due to a substantial anomaly in the development of atmospheric processes. Ordinarily, in the wintertime, the most intense cyclonic activity develops over the Norwegian and Barents Seas. Cyclones from the west bring warm At-

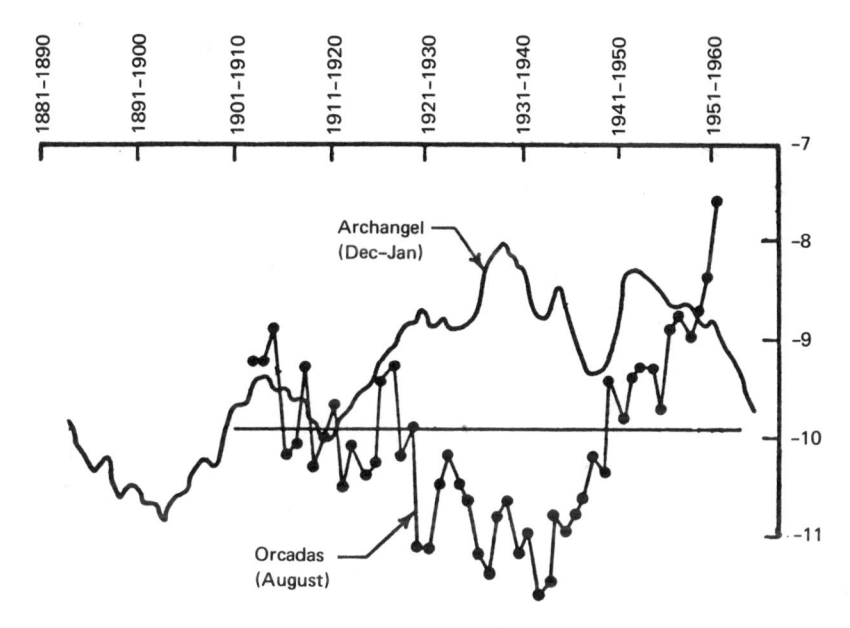

FIGURE 20 Trends of winter air temperature (°C) at Orcadas and Archangel presented as 10-year moving averages. (Adapted from Yakovlev, 1966.)

lantic air into the northern regions of the European Territory of the USSR and cause a substantial rise in the temperature there. Thaws are not infrequently observed.

In the winter of 1965-66, the cyclones from the Atlantic passed more to the south than usual: across Western Europe and the southern or central regions of the European Territory of the USSR. The penetration of the warm masses of Atlantic air also took place further south than usual. This led to a warm winter in the southern regions. . . . In the northern parts of Europe, including the Far North (in the Murmansk and Archangelsk regions), masses of cold Arctic air prevailed.

Anomalies of a similar nature . . . prevailed also in a number of previous winters—in 1940-1943 and 1955-1965. For a number of winters in the period 1919-1939, a reverse phenomenon was observed—intensive cyclonic activity in the north of the Norwegian and Barents Seas, resulting in increased influx of warm Atlantic air to northern Europe and to the western sector of the Soviet Arctic. . . .

Periods of cold winters can be noted at the end of the last century and to some extent at the beginning of the present century (1912-1918), and a prolonged period with prevailing warm winters is distinguishable in the 1920's and 1930's up to and including 1939. Then a short period of anomalously cold winters (1940-1942) follows. The new, more persistent decrease in the mean temperature of winters began with the winter of 1955-56 and continues still in the Murmansk region at the present time. It can be assumed that this process has not yet terminated.

Figure 21 shows that the trends for Archangel were also present in other sectors of the Arctic and were most pronounced during winter. In general, the trends were opposite to those for Orcadas (Figure 14). It may be noted, however, that the major change from a cooling trend to a warming trend at Orcadas occurred from 5 to 10 years earlier than the change from warming trend to cooling trend in the Arctic. This lag suggests the possibility of an oceanic teleconnection that, if understood, might be of value for forecasting climatic trends. One suggested avenue of investigation would be to compare trends of ocean temperature for various parts of the South and North Atlantic, combined with a more detailed analysis of the heat budget of the South Atlantic (Boguslavskiy, 1967).

The northern winter of 1965-1966 was also an extreme one for North America, as were the winter of 1962-1963 and those of 1940-1943. For details, the reader is referred to the several works of Willett, who describes these periods as being characterized, in the Northern Hemisphere (Willett, 1961), by

... weakening of the zonal westerlies and subtropical easterlies and strengthening of the polar easterlies ... [which is] completely typical of a strong relative trend of the general circulation towards a cellular blocking pattern going into the major maximum.

An important climatic consequence of the relatively zonal character of the general circulation during the minor in contrast to the major maximum (MM) phase of the double sunspot cycle is indicated also by the summer drought pattern. The zonal pattern should favor relatively maritime conditions, as opposed to relatively continental, in the interior of continents in middle and lower middle latitudes. All of the prolonged drought periods in our western plains have fallen in the major maximum decades, that is, in the nineties, the teens, the thirties and the fifties. The intervening decades have been free of prolonged drought.

The winter of 1962-1963 was similarly severe in Europe and North America. The anomalous conditions of January 1963 have been described by Landsberg (1963), who concludes

If it were possible for a permanent circulation pattern to establish itself in the particular fashion shown for January 1963, one could readily envision that this might be similar to those that were prevalent during the formative stages of glaciation. For this reason it would be highly desirable if it were possible to attribute causes to anomalies of a short interval of time, such as January 1963, in order to

Ice on the Ocean and World Climate 37

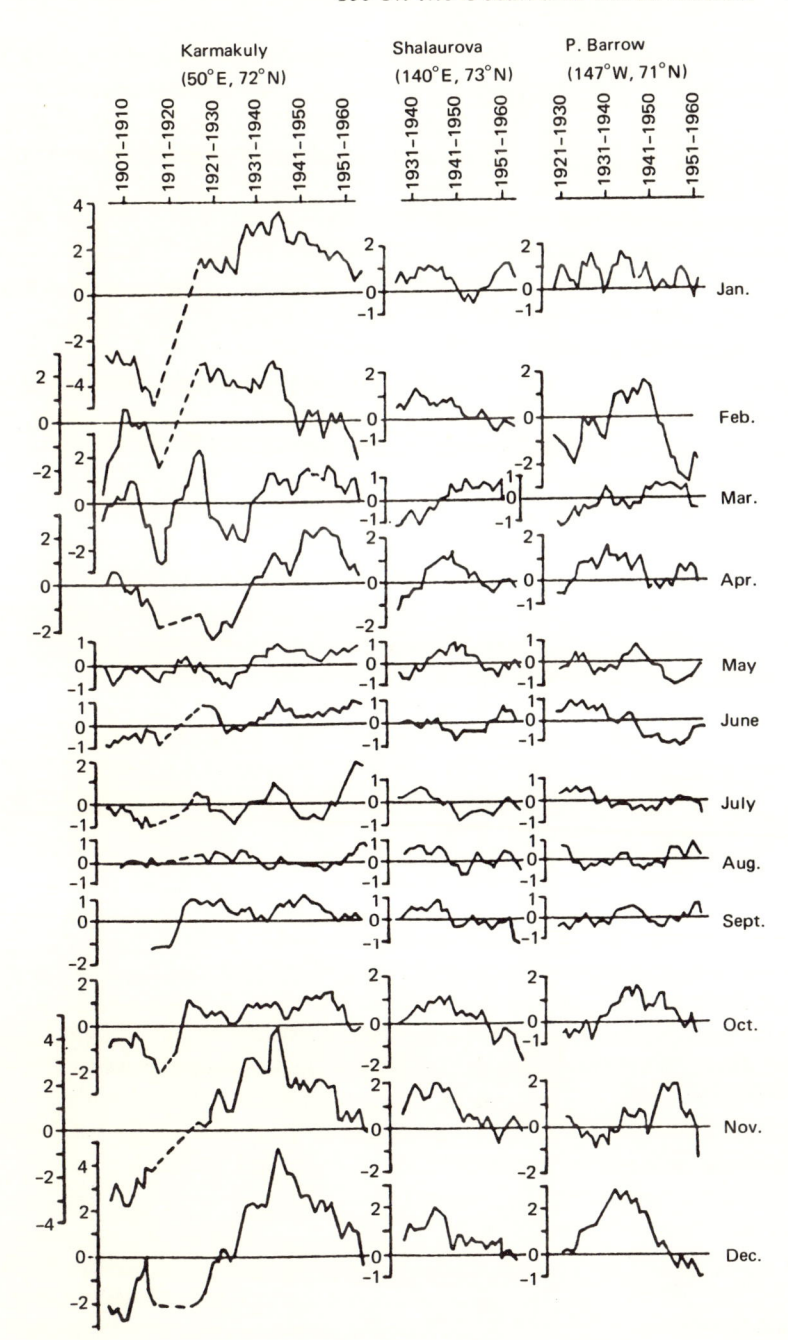

FIGURE 21 Anomalies of temperature in various sectors of the Arctic in 10-year moving averages. (Adapted from Prik, 1968.)

Iciness of Arctic Seas

In Figure 22, iciness of the Antarctic Ocean is compared with iciness of various Arctic seas. As an index of Antarctic iciness, Figure 22 shows the number of days per year that Scotia Bay was closed by ice. This index thus includes both closing and opening dates shown by previous curves. The index of general iciness of the North Atlantic is represented by the area of ice in Davis Strait (Vladimirov, 1964; Maximov, 1954). It can be seen that the trends of iciness in Davis Strait are inversely related to trends of iciness in the Weddell Sea, though there are gaps in the Davis Strait data.

From comparison with other data, it appears that the time series for Davis Strait is generally indicative of the Greenland area and the central Arctic Basin. However, some peripheral seas follow different patterns: for example, the pattern for the Sea of Okhotsk varies inversely from that for Davis Strait (Kryndin, 1964). The cause seems to be related to the prevalence of zonal (westerly-northwesterly) circulation, which carries cold air from the Siberian anticyclone over the Sea of Okhotsk. When zonal circulation is strong, there is much ice because heat loss from the ocean to the atmosphere is greater. Conversely, when a highly developed Siberian anticyclone tends to block zonal circulation, wind speeds are reduced and there is less ice in the Sea of Okhotsk.

The winter of 1942-1943 illustrates the foregoing relationships. Zonal circulation intensity was relatively weak in both hemispheres. In the Northern Hemisphere, a highly developed Siberian anticyclone contributed to a record meridional index (Figure 22). Bitter cold prevailed over the continental areas dominated by the Siberian anticyclone. In the Kamchatka area, however, 1942-1943 was a record warm winter, with an abnormally large number of cyclones, only about a third the usual number of degree days of frost, and a record minimum of sea ice.

FORECASTING FUTURE CLIMATIC TRENDS

Because the mechanics of global climate are still too poorly understood, it is not yet possible to forecast future climatic trends on the

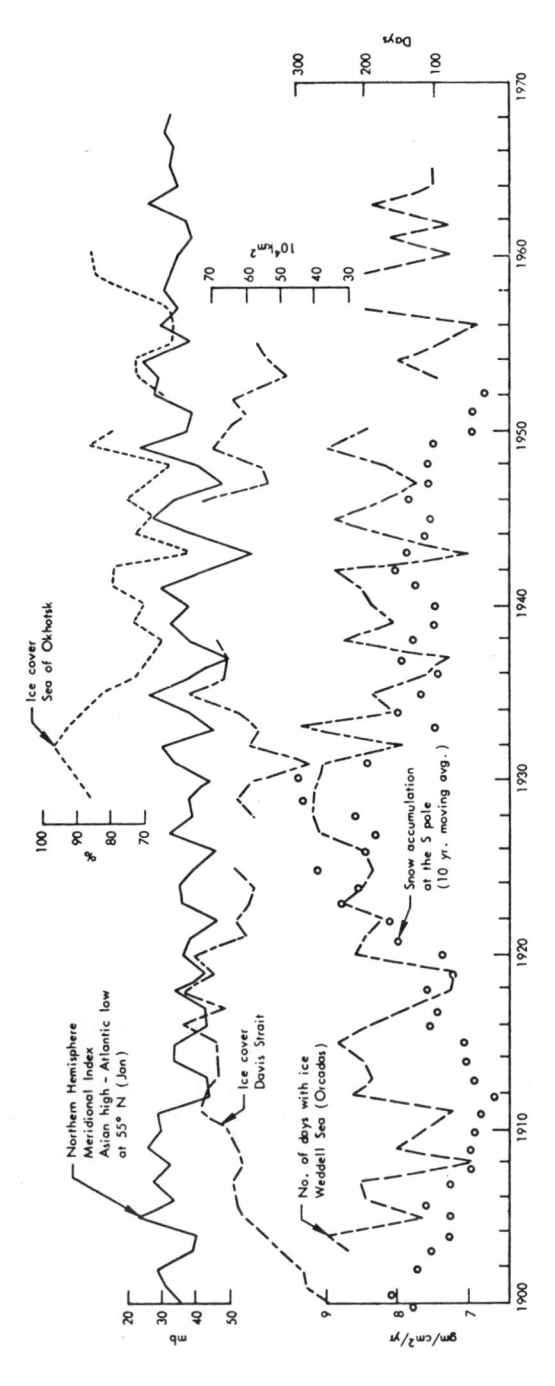

FIGURE 22 In the Antarctic region, the iciness of the Weddell Sea and the snow accumulation at the South Pole vary synchronously and can be taken as general indices of the variable intensity of the Antarctic heat sink and of atmospheric circulation intensity. North Atlantic ice extent, as represented by that of the Davis Strait, varies inversely with snow accumulation, while in the Sea of Okhotsk, it varies synchronously.

basis of established cause-and-effect relationships. Even the relative importance of such factors as atmospheric dust, cloudiness, ozone production, and carbon dioxide has not been firmly established, and many of the complicated interactions between the ocean and atmosphere are still obscure. More comprehensive observation of the behavior of the global system over a period of time will eventually clarify many of these questions, and more sophisticated mathematical models of ocean–atmosphere circulation will make it possible to simulate variations in global circulation caused by specified changes in basic causal factors. These advances will make it possible to test specific theories and should lead to rapid progress toward understanding. Meanwhile, forecasts of future climatic trends will continue to be based on the projection in time of empirical relationships between climatic changes and physical factors, such as sunspot number, whose future courses can be estimated.

For example, the warming trends early in this century were attributed by some authors to increasing contamination of the atmosphere by carbon dioxide, and there were forecasts of faster warming due to greater contamination. The observed cooling trends of recent decades, however, have shifted emphasis to theories that invoke turbidity of the atmosphere as a basic causal factor. These theories will in turn be deemphasized if the present cooling trend is followed by warming while turbidity continues to increase. In short, qualitative arguments have been quite inconclusive, and rapid progress will probably await the development of quantitative methods of observing and simulating ocean–atmosphere interactions and thermal forcing of the dynamical system.

Most investigators of climatic change believe that climatic variations are related to solar variations, even though the causal processes are not understood. Several forecasts of future climatic trends have been made by the projection of empirical relationships between solar activity and climatic variations.

Willett (1951) correctly forecast a substantial cooling over most of the Northern Hemisphere during the 1950's and 1960's. In 1961, Willett discussed the 80-to-90-year climatic cycle in more detail, and suggested that the 2 decades following the 1957 sunspot maximum would be characterized by "a low-latitude zonal pattern of essentially cool maritime climate in middle and lower latitudes."

Vitels (1962) projects lower solar activity and weakening of the planetary atmospheric circulation during the last third of this century, accompanied by greater continentality, southward displacement of cyclone paths, and colder winters with more extensive ice in the Arctic.

Girs (1956) and Girs and Dydina (1963) go into much greater detail in relating local changes in ice conditions and climate to the various types of planetary circulation. In general, Girs also forecasts for the last third of this century a weakening of interlatitudinal air exchange and a colder Arctic, especially during the descending branch of the solar cycle from 1970 to 1976.

Maximov and Smirnov (1965) forecast the frequency of each of the main circulation types discussed by Girs. For the 1970-1975 period, the most conspicuous feature of their forecast is an increase in continentality, with higher-amplitude wave disturbances in a weakening zonal flow, characteristic of "meridional circulation types." According to Girs' 1956 analysis, this forecast implied a colder Arctic, with more severe ice conditions in the Barents, East Siberian, and Chuckchi seas, and less severe ice conditions in the Kara and Laptev seas.*

For the mid-1970's to the mid-1980's, Maximov and Smirnov forecast a reverse trend, toward less continentality and a minimum of meridional circulation in the 1982-1987 period. According to Girs (1956), this trend implies a generally colder Arctic, with winds from the southern quarter somewhat facilitating ice navigation in the European Arctic and winds from the northeast quarter in the Alaskan sectors producing difficult ice conditions.

Rubinstein and Polozova (1966) present detailed data on temperature trends at stations around the planet. On the basis of an analysis of those data and the projection of observed trends, they forecast cooler winters at Salekhard, a location that should be typical of the Siberian Arctic. The reversal of the warming trend in the Greenland region (Upernavik) seems to lag considerably behind this reversal in other Arctic regions, but it may be expected to become more apparent in the coming decade.

Dzerdzeyevskiy (1968) presents the most detailed analysis of past trends of planetary circulation types and cautiously forecasts future trends, including a continuation of the trend from zonal to meridional circulation until 1973-1975.

* Californians may note that the forecast increase in M2-type circulation for the Pacific region during 1970-1975 implies a return to the wetter Southern California climate of the 1930's, characterized by a wintertime trough from the Gulf of Alaska southeast to Baja California that brings heavy and prolonged rains. The floods of January 1969 were typical of this situation, and the implication is that this trend will continue to the mid-1970's.

In summary, it appears that the consensus of empirical forecasts calls for a trend, extending to the mid-1970's, toward greater continentality and weaker zonal flow in the Northern Hemisphere. On this basis, Icelandic fishermen, for whom 1968 was the worst year in more than half a century, must look forward to more bad ice years; Baja California farmers, on the other hand, who have been suffering from sinking water tables since the 1930's, can look forward to more rain; and similar significant changes can be anticipated in other marginal climatic zones.

If, as we suggested by Figure 17, weakening zonal flow with cooling and continentality in the Northern Hemisphere is accompanied by a southward displacement of zonal flow in the Southern Hemisphere, then the northern edge of the Chilean rainfall zone should move south, bringing drought to central Chile.

These forecasts, which are based on empirical correlations, demonstrate the urgency of replacing empirical forecasts with objective numerical simulations of the planetary ocean–atmosphere system, for the social consequences of climatic variations are rapidly assuming greater importance.

REQUIREMENTS FOR FUTURE RESEARCH

Fundamental to the study of climatic change is the development of a quantitative understanding of the general circulations of the atmosphere and the oceans. Such an understanding must begin with the global distribution of heat loss and heat gain by the atmosphere and the ocean. This understanding, combined with the application of the fundamental physical laws embodied in the classical equations of motion and thermodynamics and of the conservation of mass, energy, and momentum, should allow us to predict the global distribution of temperature, pressure, and motion, together with the resulting heat and moisture transport.

It has long been the goal of meteorology to study this global thermodynamic system as a whole; however, two barriers have intervened. One has been that we have been able to observe no more than about 20 percent of the global system at any one time, so that, like the blind men examining the elephant, we have been unable to describe the whole animal. The other has been that the behavior of the system, although it obeys basically simple physical laws, is influenced by interactions so complex as to have defied quantitative analysis.

Both barriers are now crumbling. The analytical barrier is crumbling under the impact of modern computer technology, which is at last making it possible to solve mathematical representations of the global processes. Although exact and complete solutions have not yet been obtained, the rapid development of mathematical models of atmospheric and oceanic circulation is giving us new and powerful tools for investigating climate. With computers already in sight that are hundreds of times more powerful than present models, rapid progress toward a quantitative understanding of global climate can be expected.

The observational barrier is crumbling primarily under the impact of satellite observation systems. A single polar-orbiting satellite can survey the entire earth–atmosphere system every day. Satellites have thus far been used to infer the state of the ocean–atmosphere system mostly through observation of cloud patterns, but the use of quantitative sensors to monitor important heat-exchange patterns is increasing.

The line of investigation proposed here might be summarized as follows:

1. Obtain a quantitative understanding of the planetary heat budget and the physical processes controlling it.
2. Observe and explain the relationships between variations in heat budget and variations in atmospheric and oceanic circulation.
3. Demonstrate these relationships experimentally by mathematically modeling the dynamic behavior of the atmosphere and oceans.
4. Assess ways of purposefully influencing large-scale processes—and test these schemes on models.

To provide input to our dynamical models, we are most interested in monitoring, over the whole planet, the heat gain and heat loss by the atmosphere and by the ocean. This basically entails the measurement of the heat balance at the earth's surface, which satellites may some day be able to do. In addition, an enormous amount of field observation must be conducted by ships, aircraft, balloons, and surface stations, in all seasons, to understand and measure the important heat-exchange processes. Only then will we be able to interpret satellite observations adequately.

I believe that the most important factors to monitor in both the Arctic and Antarctic are the seasonal patterns of surface heat exchange over the oceanic regions. In particular, we must know the location and extent of pack ice, open leads, and polynyas, especially in winter, and the variation of mean areal albedo during summer. Continuing monitoring programs should be carried out that will give us permanent and reliable data about year-to-year variations.

PROSPECTS OF INFLUENCING PACK-ICE EXTENT AND CLIMATE

Since the thickness and extent of pack ice is sensitive to various heat-budget factors, the possibility of influencing climate by changing the amount of pack ice deserves consideration.

If it were possible to influence the extent of Antarctic pack ice materially, it might also be possible to exaggerate or to diminish the natural trends of global climate. Changing the ice cover in the Arctic, on the other hand, would afford a more local, and perhaps a more effective, means of influencing climate in northern latitudes, and if an ice-free ocean would tend to remain ice-free, as Professor Donn suggests, then changing the Arctic ice cover might be much more lasting and more economical. In fact, the possibility of entirely removing the Arctic pack ice has been under discussion for many years (Rodahl, 1953). Various schemes have been suggested; a Bering Strait dam, for example, was proposed as a way to induce a greater influx of oceanic heat into the Arctic Basin (Borisov, 1959, 1967).

The use of nuclear explosives has been proposed as a possible way to create a more turbid winter atmosphere and thus to reduce radiative loss from the surface (Wexler, 1958). Destruction of the insulating snow cover by waves induced by nuclear explosives has been suggested as a possible way either to increase ice production in winter or to hasten ice melting in summer (Fletcher, 1965). Many authors have suggested dusting the surface of the ice with a dark substance to increase solar absorption; others have dismissed the idea as logistically unfeasible.

With regard to these various schemes, I would like to make two observations. One is that, however objectionable some of these schemes may be for political, economic, or environmental reasons, there is good reason to believe that we probably do possess the engineering capability to influence the extent of, or even eliminate, the ice should we really need to do so. We noted previously, for example (Figure 5), that advancing the date of puddling by only three or four weeks would make an enormous difference in the heat budget, and

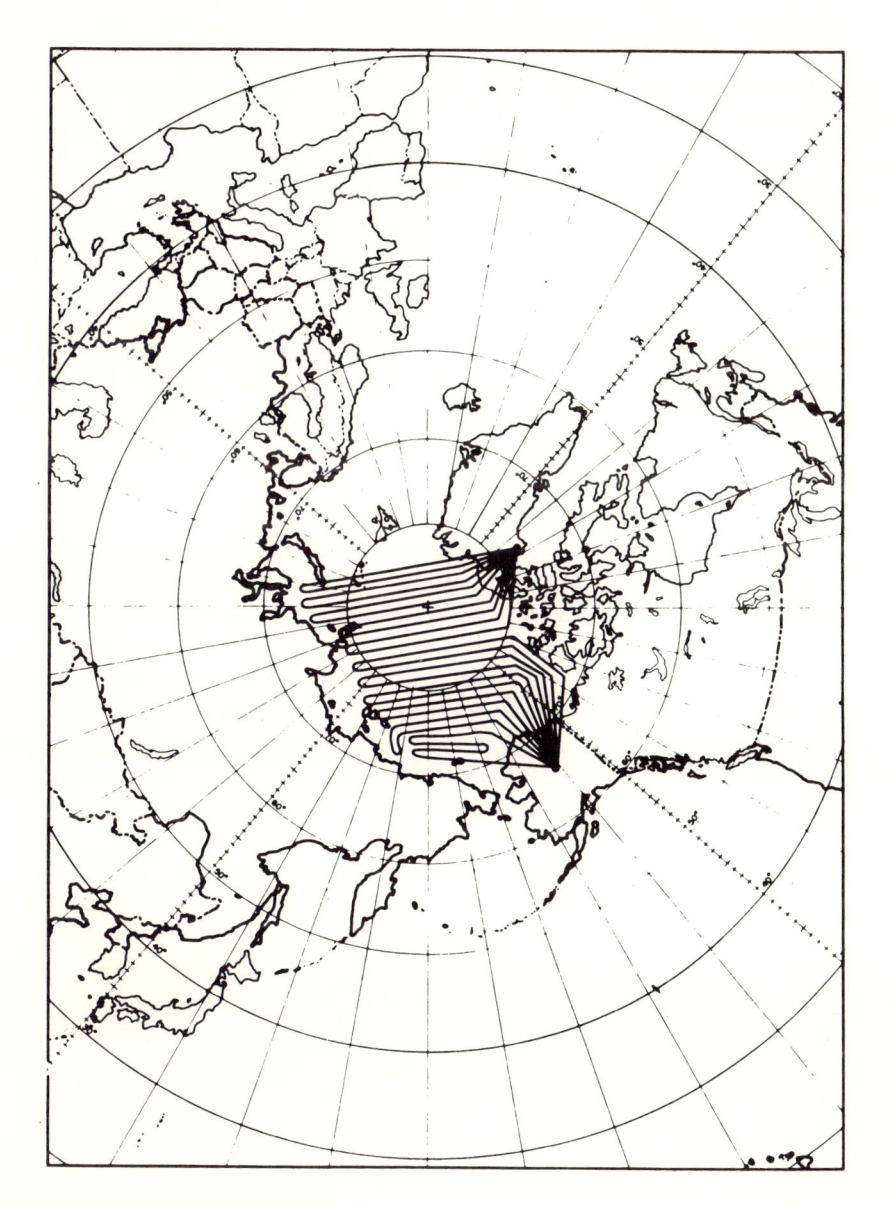

FIGURE 23 Possible flight paths over the Arctic pack ice.

when we examine the factors that control the heat-budget components, we find that some of them are very sensitive to small changes. The creation of thin, high cirrus, for example, greatly reduces the flow of long-wave radiation to space without greatly reducing incoming solar radiation. Under certain conditions, the dispersing of relatively small amounts of reagents can dissipate stratus clouds (with a high surface albedo, this actually reduces melting, but with low surface albedo, it hastens melting). Under some conditions, it may even be worthwhile to consider dispersing a dark powder to lower the surface albedo in local areas.

To put these possibilities into operational perspective, Figure 23 shows a system of flight profiles over the Arctic Basin from Eielson and Thule Air Force Bases. There are 11 flight profiles, each representing an average flight time of about 6 hours. At 12 hours per day of aircraft utilization, which is not unrealistic for a shuttle operation, all of these profiles could be reflown every day with six aircraft. The legs as drawn are about 60 miles apart; 60 aircraft could thus permit profiles one mile apart and cover the area every day. Suppose we do it for only 30 days, say, from mid-April to mid-May, when solar radiation is rapidly increasing; in 30 days we will have flown 30 profiles to the mile, and if a fleet of aircraft like the C-5 military transports were used, we could disperse about 300,000 tons of material. Such an effort might well be effective, but it is difficult to make quantitative estimates because all of the unknown factors have not been evaluated. For example, how thick a dusting substance should be laid on? And how much would it reduce the albedo over an area of a particular size? A fleet of C-5's is not needed to investigate these questions—even an old broken-down DC-3 will do. Nature repeats her cycle only once a year; the sooner we get started, the sooner we will learn the answers to such questions.

The primary factor limiting our ability to influence climate is our poor understanding of how our climatic machine operates—no matter what we are physically capable of doing. We cannot apply our technological capabilities in a purposeful way until we can predict the consequences for the entire ocean and atmosphere, which is a single, interacting, physical system.

REFERENCES

Ahlmann, H. W., 1945, "The Organization of Soviet Arctic Research," in *Polarboken,* Stockholm, Sweden, 32.

"Atlas of Antarctica," *Soviet Geography*, May–June 1967 (translation by the American Geographical Society from "Atlas of the Antarctic," Vol. 1, Moscow, 1966).

Badgley, F., 1961, "Heat Balance at the Surface of the Arctic Ocean," *Proc. Western Snow Conference*, Spokane, Washington. (Unpublished.)

Barkov, N. I., and V. N. Petrov, 1966, "Variations of Snow Accumulation and Some Peculiarities of the Climate and Atmosphere Circulation of Antarctica," *Soviet Antarctic Expedition, Bulletin 59*.

Boguslavskiy, S. G., 1967, "Determination of the Discharge of the Antarctic Convergence in the Atlantic Ocean," *Tr. Morskogo Gidrofiz Inst., 38,* 36–41.

Borisenkov, Ye P., 1965, "Energy Transformation in the Atmosphere of the Northern and Southern Hemispheres," *Meteorol. Issledovaniya, 9,* 5–13 (in *NASA TT-F-396*).

Borisov, P. M., 1959, "The Bering Strait Dam," *Lit. Gaz., 24.*

Borisov, P. M., 1967, "Can We Control the Climate of the Arctic?" *Priroda*, Issue 12, Moscow, 63–73.

Brooks, C. E. P., 1949, *Climate Through the Ages*, McGraw-Hill, New York.

Budyko, M. I., 1961, "The Thermal Zonality of the Earth," *Meteorol. Gidrol., No. 11,* 7–14.

Budyko, M. I., 1962, "Polar Ice and Climate," *Izv. Akad. Nauk SSSR*, Ser. Geogr. No. 6.

Budyko, M. I., 1964, *Atlas of the Heat Balance of the Earth*, Hydrometeorological Publishing House, Leningrad.

Budyko, M. I., 1967, "Variation of Climate," *Meteorol. Gidrol., No. 11,* 18–27.

Budyko, M. I., 1968, "The Effect of Solar Radiation Variations on the Climate of the Earth," *Proceedings of the International Radiation Symposium*, Bergen, Norway.

Dolgina, I. M. (ed.), 1967, *Aeroclimatic Handbook of Antarctica, 2*, Hydrometeorological Publishing House, Leningrad.

Donn, W. L., and D. M. Shaw, 1965, "The Stability of an Ice Free Arctic Ocean," *J. Geophys. Res., 71,* 1086–1095.

Dzerdzeyevskiy, B. L., 1968, "Climatic Fluctuations and the Problem of a Super Long Range Forecast," *Izv. Akad. Nauk SSSR*, Ser. Geogr., No. 5.

Fletcher, J. O., 1965, "The Heat Budget of the Arctic Basin and Its Relation to Climate," *R-444-PR*, The Rand Corporation, Santa Monica, California.

Giovinetto, M. B., and W. Schwerdtfeger, 1966, "Analysis of a 200 Year Snow Accumulation Series from the South Pole," *Arch. Meteorol., Geophys. Bioklimatol., 15, 2.*

Girs, A. A., 1956, *Interrelation of Processes in the Atmosphere and Hydrosphere*, OPNAV Publ. P03-31.

Girs, A. A., and L. A. Dydina [eds.], 1963, "Contributions to Long-Range Weather Forecasting for the Arctic," *Tr. Arkt. Antarkt. Inst., 225,* Leningrad (Translated by National Science Foundation, 1966).

Kotlyakov, V. M., 1962, "Change in Snow Accumulation in Antarctica in the Last Ten Years," *Soviet Antarctic Expedition, Bulletin No. 35*.

Kryndin, A. N., 1964, "Seasonal and Inter-Annual Variations in the Ice Conditions and the Position of the Ice Boundaries of the Far-Eastern Seas in Relation to Peculiarities of Atmospheric Circulation," *Tr. Gosudarstvennogo Okeanogr. Inst., 71,* Moscow, 5–82.

Laikhtman, D. L., 1959, "Certain Principles of the Heat Regime in the Central Arctic," *Tr. Arkt. Antarkt. Inst., 226,* Leningrad.

Lamb, H. H., 1966, *The Changing Climate,* Methuen, London, 1966.

Lamb, H. H., and A. I. Johnson, 1966, "Secular Variations of the Atmospheric Circulation Since 1750," in *Geophysical Memoir XIV,* 110.

Maximov, I. V., 1954, "Secular Fluctuations of Ice Conditions in the Northern Part of the Atlantic Ocean," *Tr. Inst. Okeanol., 8.*

Maximov, I. V., and N. P. Smirnov, 1965, "Experiment at Constructing a Long-Range Forecast of the Basic Forms of Atmospheric Circulation in the Northern Hemisphere by Harmonic Analysis," *Tr. Arkt. Antarkt. Inst., 262.*

Mitchell, J. M., Jr., 1963, "On the World Wide Pattern of Secular Temperature Change," in *Changes of Climate,* UNESCO, Paris, France.

Nazarov, V. S., 1962, "Ice of the Antarctic Waters—Results of Research of the IGY Program," *Okeanologiya, 6.*

Petrov, V. N., 1964, "Variations of Snow Accumulation in Antarctica from 1891 to 1955," *Soviet Antarctic Expedition, Bulletin 47.*

Prik, Z. M., 1968, "On the Fluctuations of the Climate of the Arctic and the Reasons for Them," *Tr. Arkt. Antarkt. Inst., 274.*

Privett, D. W., 1960, "The Exchange of Energy Between the Atmosphere and the Oceans of the Southern Hemisphere," *Geophysical Memoir CIV.*

Putnins, P., 1963, *Studies on the Meteorology of Greenland,* U.S. Weather Bureau, Washington, D.C.

Rakipova, O. R., 1966, "The Influence of Arctic Ice Cover on the Zonal Distribution of Atmospheric Temperature," *in* Proceedings of the Symposium on Arctic Heat Budget and Atmospheric Circulation. Rand Corporation publication, RAND-RM-5233-NSF, pp. 411-441.

Rodahl, K., 1953, *North,* Harper, New York.

Rubinstein, E. S., and L. G. Polozova, 1966, *Sovremennoe Izmeneniye Klimata,* Hydrometeorological Publishing House, Leningrad.

Sawyer, J. S., 1966, "Possible Variations of the General Circulation of the Atmosphere," in *World Climate from 8000 B.C. to 0 B.C.,* Royal Meteorological Society, London.

Schwerdtfeger, W., 1956, "The Semi-Annual Pressure Oscillation, Its Cause and Effects," *J. Meteorol., 13,* 217-218.

Schwerdtfeger, W., 1959, *Meteorologia Descriptira del Sector Antartico Sudamericano,* Instituto Antartico Argentino, Publication No. 7.

Treshnikov, A. F., 1967, "The Ice of the Southern Ocean," *Proceedings Symposium on Pacific-Antarctic Sciences,* Eleventh Pacific Science Congress, Tokyo.

Untersteiner, N., 1964, "Calculations of Temperature Regime and Heat Budget of Sea Ice in the Central Arctic," *J. Geophys. Res., 69.*

Vitels, L. A., 1962, "Anomalies of the Cyclic Variations of Solar Activity and the Tendency of Recent Climatic Variations," *Tr. Gl. Geofiz. Obzerv., 133,* 35-54.

Vladimirov, D. A., 1964, "Ice Conditions in Davis Strait," *Tr. Gos. Okeanogr., 71,* 100-119.

Wexler, H., 1958, "Modifying Weather on a Large Scale," *Science, 128,* 1059-1063.

Willett, H. C., 1951, "Extrapolation of Sunspot-Climate Relationships," *J. Meteorol., 8,* 1-7.

Willett, H. C., 1953, "Atmospheric and Ocean Circulation as Factors in Glacial-Interglacial Changes of Climate," in *Climatic Change,* Harvard University Press, Cambridge, Mass.

Willett, H. C., 1961, "Solar Climatic Relationships," *Ann. N.Y. Acad. Sci., 95,* 89-106.

Willett, H. C., 1964, "Evidence of Solar-Climatic Relationships," *Symposium on Weather and Our Food Supply,* University of Iowa, Ames.

Willett, H. C., 1965, "Solar Climatic Relationships in the Light of Standardized Climatic Data," *J. Atmos. Sci., 22,* 120-136.

Yakovlev, B. A., 1966, "An Unusually Cold Winter in the Murmansk Region," in *Chelovek i Stikhiya* (*Man and the Elements*), Hydrometeorological Publishing House, Leningrad.

DISCUSSION

William L. Donn and *David M. Shaw*
LAMONT-DOHERTY GEOLOGICAL OBSERVATORY

The present north polar sea has a perennial ice cover. During the summer, the uniformity of this cover is broken by occasional leads (narrow channels of open water). The surface becomes somewhat softened and "slushy" from the influence of the continuous, albeit weak, sunlight. However, despite these seasonal variations in the quality of the surface, the very presence of an essentially permanent cover plays a vastly important role in determining the climate of the Arctic region.

The effects of snow cover on midlatitude climate are often obvious when deviations from the normal occur. Namias (1963) demonstrated that an extended snow cover over the central United States from mid-February to mid-March 1960 produced a temperature anomaly of 6°C below normal in this region. A similar, and perhaps stronger, effect occurred in the winter of 1962-1963 in both North America and Eurasia. Adem (1965) showed that the December 1962 snow cover extended several degrees farther south than was normal for the northern United States and large parts of Europe and Asia. Temperatures in North America and Eurasia reached 6°C to 10°C below normal, respectively, during the following month, January 1963. The reason for this effect lies in the high albedo of snow, which, when fresh, has a value of 80 to 90 percent, resulting in a profound reduction in the absorption of sunlight.

The north polar sea ice does not fluctuate drastically, but we may try to anticipate climatic changes that might occur if the ice cover were to disappear. Recent reports referring to a shrinkage of the Arctic ice pack (e.g., Zubov, 1948; Ahlmann, 1953; Petterssen, 1964) lend urgency to such climatic prognosis. An experiment by Arnold (1961) has shown that summer dusting of the Arctic ice with soot can provide sufficient additional absorption of solar energy to melt the soot-dusted ice completely. Other, more popular, reports have suggested ice removal by a complete dusting of the Arctic ice pack or by thermonuclear processes in order to improve the climate of the Arctic region. The possibility of increasing the "greenhouse effect" by appropriate aerosols has also been considered. Ewing and Donn (1956), and Donn and Ewing (1966) have theorized that rather than promote a more salubrious polar climate, an ice-free Arctic would trigger a new glacial stage by providing moisture for snowfall over the presently arid polar lands.

In order to evaluate the effects of an ice-free Arctic Ocean, it is first necessary to determine whether an ice-free Arctic can be maintained under the present solar radiation regime in the high latitudes. That is the chief purpose of this paper.

ARCTIC ALBEDO: WITH AND WITHOUT ICE

Of the factors involved in the Arctic heat budget, the surface albedo is of primary importance. The albedo of the present surface is rather variable, changing from winter to summer. Although Fritz (1959) reported a value of 75 percent in August 1955, the values normally estimated are lower. According to Larsson and Orvig (1961), the albedo varies from about 70 percent in spring and fall to 50 percent in summer. A weighted mean of 61 percent has been adopted by us for the Arctic sea-ice albedo during the solar radiation seasons. Figures 1 and 2 indicate the changes in Arctic Ocean albedo from winter to summer.

In Figures 1 and 2, areas with an albedo of less than 20 percent represent open ocean. With the removal of the polar sea ice, this con-

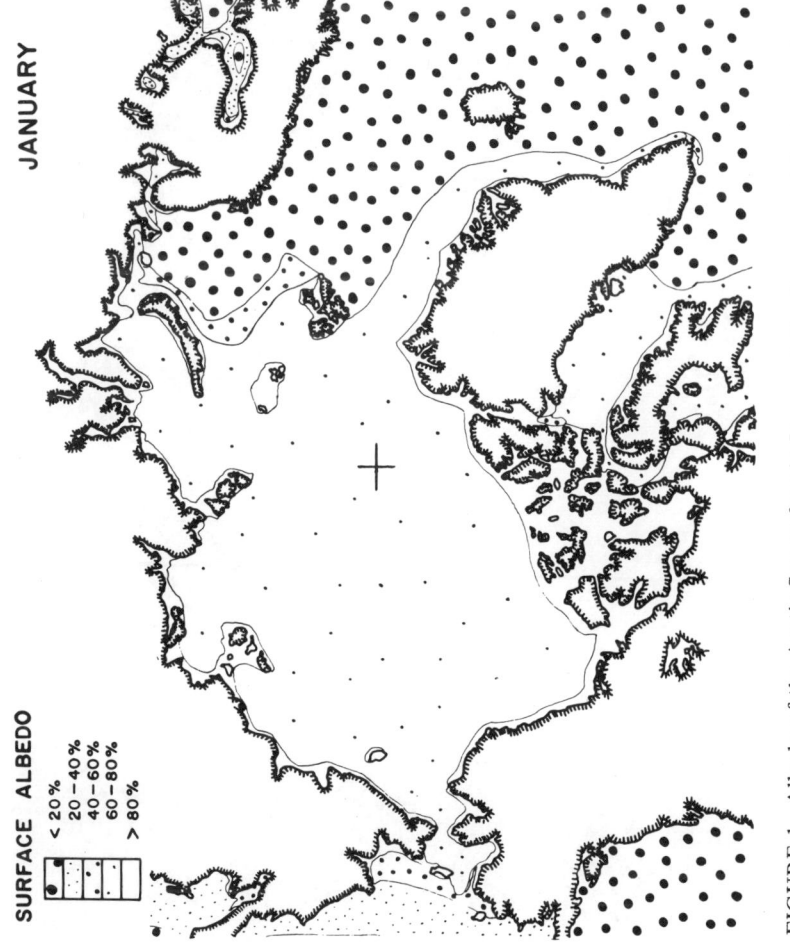

FIGURE 1 Albedo of the Arctic Sea surface in January (after Larsson and Orvig, 1961).

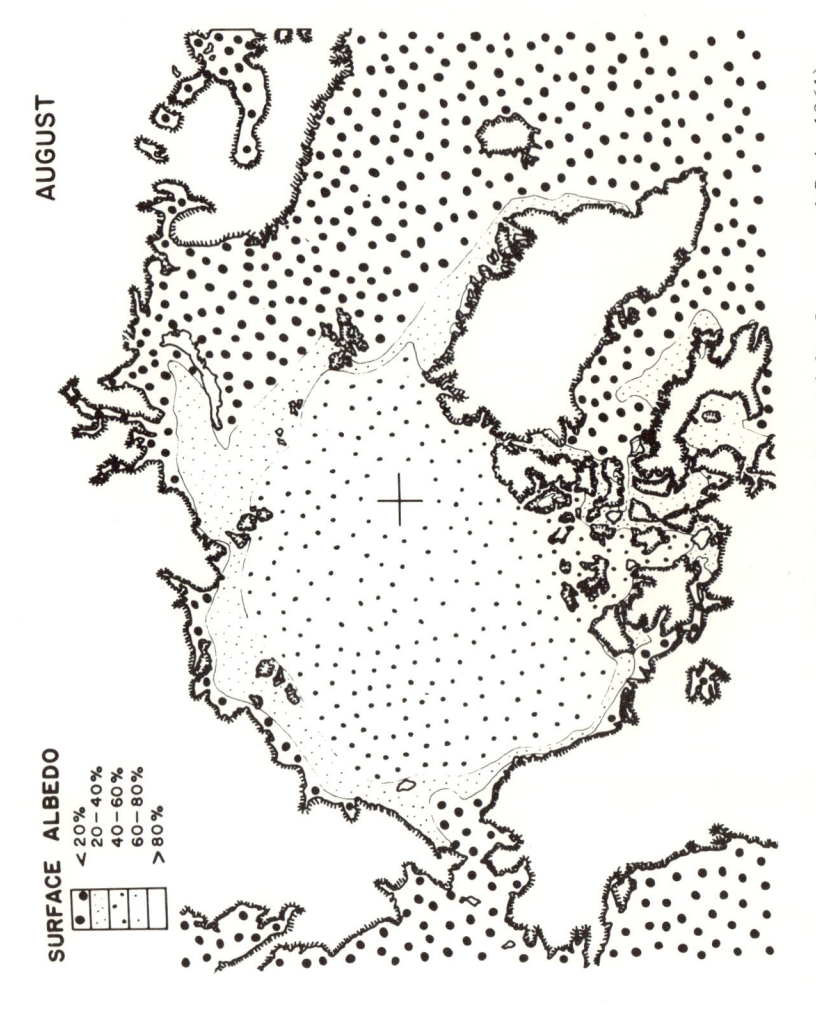

FIGURE 2 Albedo of the Arctic Sea surface in August (after Larsson and Orvig, 1961).

dition would extend through the entire polar region, which is at present characterized by high albedo values. The actual value is, of course, much lower than 20 percent. For diffuse solar radiation, Powell and Clarke (1936) give a value of 6.5 percent for an ocean albedo. Houghton (1954) determined a value of 8.3 percent for latitude 60 degrees on the basis of weighted values for radiation under clear and cloudy skies. For latitude 70 degrees, Budyko (1956) gives a mean summer albedo of 9 percent. In our heat-budget calculations, given below, we used a conservative value of 10 percent for an ice-free ocean, in view of the higher average latitudes involved.

HEAT BALANCE OF THE ICE-COVERED POLAR SEA

Heat balance calculations for an ice-covered and an ice-free polar sea and for the associated sea–atmosphere system have been carried out by Donn and Shaw (1966) on the basis of the above albedo values. The heat sources considered were short-wave solar radiation and ocean-current heat flux, of which the first is by far the more important. Radiation data from ice island T-3 and the Soviet ice-floe station NP-2 were utilized in estimating the net radiation absorption over the entire polar sea, an area of 10^{17} cm^2. The heat sinks considered were surface long-wave radiation, latent heat of vaporization, and sensible heat. Radiation loss was computed with the use of the Elsasser radiation diagram and the appropriate values of mean surface temperature, atmospheric humidity, and cloud cover.

Results of the calculations are summarized in Table 1, which gives the heat balance for an entire year for an ice-covered polar sea (10^{17} cm^2). Although the heat balance value of 0.01 × 10^{21} cal for the

TABLE 1 Annual Heat Balance of Ice-Covered Polar Sea (area, 10^{17} cm^2), in 10^{21} cal

Heat Source	Gain	Loss
Net short-wave radiation absorbed	2.83	–
Ocean-current heat flux	0.16	–
Long-wave radiation	–	2.29
Latent heat (evaporation)	–	0.31
Sensible heat	–	0.38
Balance	0.01	

year is smaller than the probable error in the factors involved in its determination, it is quite close to that obtained by several other investigators. But, more important, the results indicate that the heat balance of the Arctic sea surface is very close to equilibrium. Hence, a significant change in any of the factors involved would drive the thermal balance strongly positive or negative. The removal of the ice would surely provide the means for such a change, as a consequence of the drop in albedo.

HEAT BALANCE OF AN ICE-FREE POLAR SEA

With an open, ice-free polar sea, the albedo decrease from 61 to 10 percent would result in an increase in the heat content of surface waters from absorption of short-wave radiation. Also, some increase in the water interchange between the Arctic and Atlantic oceans should occur under ice-free conditions, again increasing heat input to the Arctic. However, this input of 0.32×10^{21} cal/yr is small, compared with the increase in absorption of short-wave radiation of 5.24×10^{21} cal/yr. Heat sink values used in the calculations are those for ice-free water at $0°C$. A pronounced decrease in long-wave radiation loss would occur because of the increased greenhouse effect of the moister and somewhat cloudier skies above the open ocean. A list of the parameters contributing to the heat-balance results is shown in Table 2.

The heat balance of 2.63×10^{21} cal/yr for the ice-free polar sea (of 10^{17} cm^2 area) is more than ten times the amount of heat required to melt the entire Arctic ice pack in one year. Even a relatively large change in some of the quantities used in these calculations could

TABLE 2 Annual Heat Balance of Ice-Free Polar Sea (area, 10^{17} cm^2), in 10^{21} cal

Heat Source	Gain	Loss
Net short-wave radiation absorbed	5.24	—
Ocean-current heat flux	0.32	—
Long-wave radiation	—	0.96
Latent and sensible heat	—	1.97
Balance	2.63	

leave a significant positive heat balance for the surface, leading to the conclusion that the north polar sea would remain ice-free if the existing ice were to disappear. Our high value for the heat balance indicates that the assumed water temperature of 0°C is too low and that warming would occur until an equilibrium heat balance were achieved.

THE THERMODYNAMIC MODEL

Through the use of a recent thermodynamic model developed by Adem (1965), it is now possible to integrate the factors involved in the radiational heat balance over the entire Northern Hemisphere for ice-free and ice-covered states of the polar sea, in order to arrive at the distribution of actual temperatures over the hemisphere.

Adem's model predicts thermal states of the earth's surface and midtroposphere as a function of the heat-balance components and of horizontal mixing. Input data consist of solar radiation, surface albedo and radiation, latent heat lost from the surface and released in cloud condensation, sensible heat loss from the surface, and sea surface and midtroposphere temperatures. For given inputs at 512 grid points, a computer program solves Adem's equations, yielding the temperatures necessary to achieve equilibrium between the heat sources and sinks. A more complete explanation of the entire procedure was prepared by Shaw (1969).

Before computing the changes consequent upon an ice-free polar sea, the procedure was tested for present conditions. With the use of the best data available to us for the present hemispheric conditions, a winter mean surface temperature of −27°C was obtained for the central Arctic. The form of the results is given in Figure 3. According to the literature, an average value of −30°C has been estimated for the same region. Our result of −27°C compares well with the observed average, especially since the observed value is based on rather limited observations. Comparisons for a summer mean surface temperature are more divergent, but for understandable reasons. Our computations give a summertime mean of about +4°C for the central Arctic, compared with an observed mean of about −13°C. However, the present program does not include any way of calculating the effects of the inversion present over the polar ice. This inversion tends to

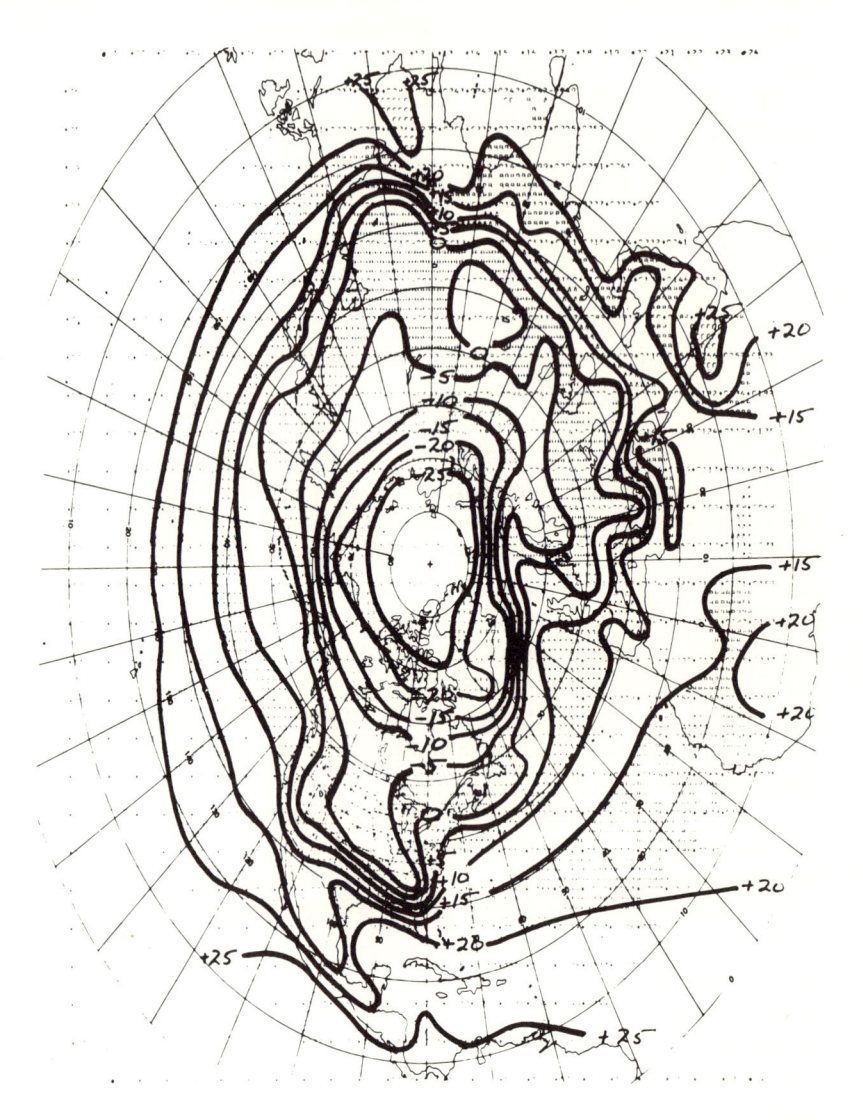

FIGURE 3 Surface temperature distribution calculated for the Northern Hemisphere winter with an ice-covered Arctic Ocean.

maintain the low-level air at the same temperature as the ice, resulting in the present low observed mean. Another comparison lies in the computed value of 12.5°C for the hemispheric mean and an observed mean of about 15°C.

Arctic mean surface air temperatures were then determined for ice-free conditions in summer and winter. Input conditions were as follows:

Albedo: 10 percent
Cloud cover: 95 percent
Heat loss by evaporation: 34 langleys/day
Sensible heat loss: 20 langleys/day
Heat of condensation in clouds: 34 langleys/day
Solar radiation: mean June insolation

The computational results for this model are shown in Figure 4, which contains 5°C isotherms drawn directly on the machine plot of Northern Hemisphere air temperature distribution. Computed air temperatures at the polar sea surface range from 0° to 2°C, despite an input of 0°C for the ice-free water.

Computations of winter surface air temperatures are based on inputs similar to those of the summer, but with no incoming radiation and a temperature of 5°C assumed on the basis of warming over the entire summer season. The results, in Figure 5, indicate an equilibrium air temperature of about 2.5°C.

These results are all preliminary, but they seem encouraging in view of the results of Budyko (1962), who concluded that polar air temperatures over an ice-free Arctic would be 10°C in summer and 5°C in winter. Less optimistic results were obtained by Fletcher (1965) and Badgley (1961), but they used an ice-free model without modifying the atmospheric model to include an increased greenhouse effect. In using the present atmosphere rather than one with a larger greenhouse effect, much greater heat loss resulted. We believe this to be unrealistic.

In continuing this analysis, we plan to carry out temperature computations on a monthly rather than a seasonal basis and to use the results for each month as one of the inputs for the following month. Regardless of the final results, we must regard the Arctic as a region of great climatic sensitivity, where climatic research should be focused before any attempts are made to change the environment artificially.

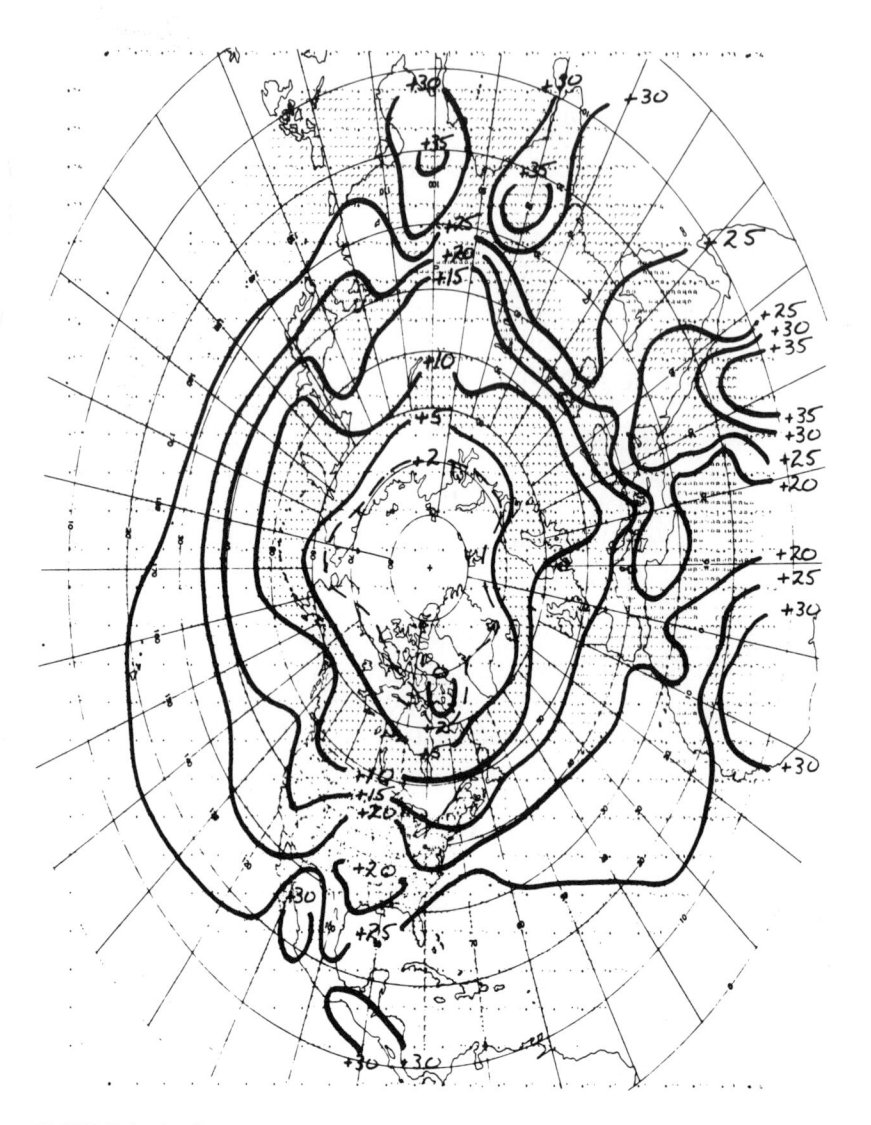

FIGURE 4 Surface temperature distribution calculated for the Northern Hemisphere summer with an ice-free Arctic Ocean.

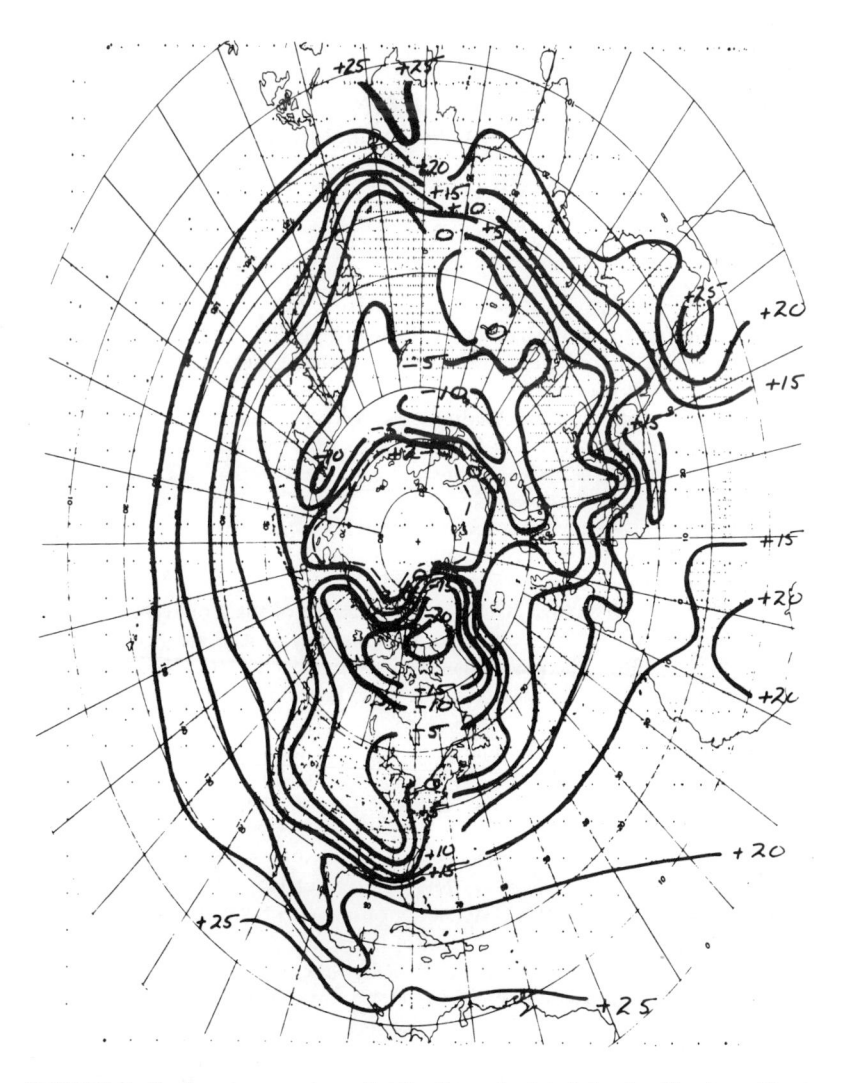

FIGURE 5 Surface temperature distribution calculated for the Northern Hemisphere winter with an ice-free Arctic Ocean.

ACKNOWLEDGMENTS

The research described was carried out under a gift from the U.S. Steel Foundation and a grant from the National Aeronautics and Space Administration [Number NsG 445(E)]. Numerical solutions of the thermodynamic model were performed on the IBM 360/90 at the Institute for Space Studies (NASA), New York City. We are grateful to J. Adem for his cooperation and for making available his thermodynamic model and the basic program used in its solution.

REFERENCES

Adem, J., 1964, "On the Physical Basis for the Numerical Prediction of Monthly and Seasonal Temperatures in the Troposphere-Ocean-Continent System," *Monthly Weather Rev., 92,* 91.

Adem, J., 1965, "Experiments Aiming at Monthly and Seasonal Numerical Weather Prediction," *Monthly Weather Rev., 93,* 495-503.

Ahlmann, J., 1953, "Glacier Variations and Climatic Fluctuations," *Am. Geogr. Soc. Bowman Mem. Lectures, Ser. 3.*

Arnold, K., 1961, "An Investigation into Methods of Accelerating the Melting of Ice and Snow by Artificial Dusting," in *Geology of the Arctic,* G. O. Raasch, [ed.], University of Toronto Press, 989-1013.

Badgley, F., 1961, "The Heat Balance at the Surface of the Arctic Ocean," *Proc. Western Snow Conf.,* Spokane, Washington.

Budyko, M. I., 1956, "The Heat Balance of the Earth's Surface (U.S. Department of Commerce translation, Washington, D.C.) Hydrometeorological Publishing House, Leningrad.

Budyko, M. I., 1962, "Polar Ice and Climate," *Izv. Akad. Nauk SSSR,* Ser. Geogr. No. 6.

Donn, W., and M. Ewing, 1966, "A Theory of Ice Ages III," *Science, 152,* 1706-1712.

Donn, W., and D. Shaw, 1966, "The Heat Budgets of an Ice-Covered and Ice-Free Arctic Ocean," *J. Geophys. Res., 71,* 1087-1093.

Ewing, M., and W. Donn, 1956. "A Theory of Ice Ages," *Science, 123,* 1061-1066.

Fletcher, J. O., 1965, "The Heat Budget of the Arctic Basin and its Relation to Climate," *The Rand Corporation Publ. R-444-PR,* Santa Monica, California.

Fritz, S., 1959, "Solar Radiation Measurements in the Arctic Ocean," in *Scientific Studies at Fletchers Ice Island, T-3* (1952-1955), 2, 7-10; *Geophys. Res. Papers* (U.S.), *No. 63.*

Houghton, H., 1954, "On the Annual Heat Balance of the Northern Hemisphere," *J. Meteorol., 11,* 1-9.

Larsson, P., and S. Orvig, 1961, "Atlas of Mean Monthly Albedo of Arctic Surfaces," *McGill Univ. Arctic Meteorol. Res. Group, Publ. 45* (Air Force Cambridge Res. Lab., 1051).

Namias, J., 1963, "Surface-Atmosphere Interactions as Fundamental Causes of

Drought and Other Climatic Fluctuations," in *Changes of Climate,* UNESCO, Paris, 345.

Petterssen, S., 1964, "Meteorological Problems: Weather Modification and Long-Range Forecasting," *Bull. Am. Meteorol. Soc., 45,* 2-11.

Powell, W., and G. Clarke, 1936, "Reflection and Absorption of Radiation at the Surface of the Ocean," *J. Opt. Soc. Am., 26,* 111-120.

Shaw, D., 1969, "Thermodynamic Studies of Paleo climates," Ph.D. Dissertation, Lamont Geological Observatory (Columbia University Library).

Zubov, N., 1948, "Arctic Ice and the Warming of the Arctic," *V. Tsentre Arktiki,* E. Hope, trans. (Defense Research Board, Ottawa, 1950).

DISCUSSION

Sigmund Fritz

NATIONAL ENVIRONMENTAL SATELLITE SERVICE

To assess the influence of ice in polar regions on weather and climate, it is important to divide the discussion according to time and space scales. For short time intervals and short distances, the presence of ice will influence the temperature in neighboring areas. For example, if after a severe winter the Arctic ice near Iceland is more extensive than normal, then with the approach of spring we would expect the air temperatures in the vicinity of Iceland to be below normal for a period of days. (For such short-period forecasts, satellite data may be helpful. Figure 1 is a picture of ice, snow, and cloud cover in the vicinity of iceland.) However, if we consider much longer time periods, then, as Fletcher mentioned, the waxing or waning of the ice will depend in part on the global circulation of the atmosphere as a whole, and the circulation in the Arctic may be influenced more by events at some distant place—such as the tropics—than it would be by the ice, although important interactions could doubtless occur.

Thus, if we go to longer time scales, we need to consider competing ideas about climatic changes and those factors that may influence circulation over the globe as a whole. This question was discussed by Fletcher (1966) and by Mitchell (1966).

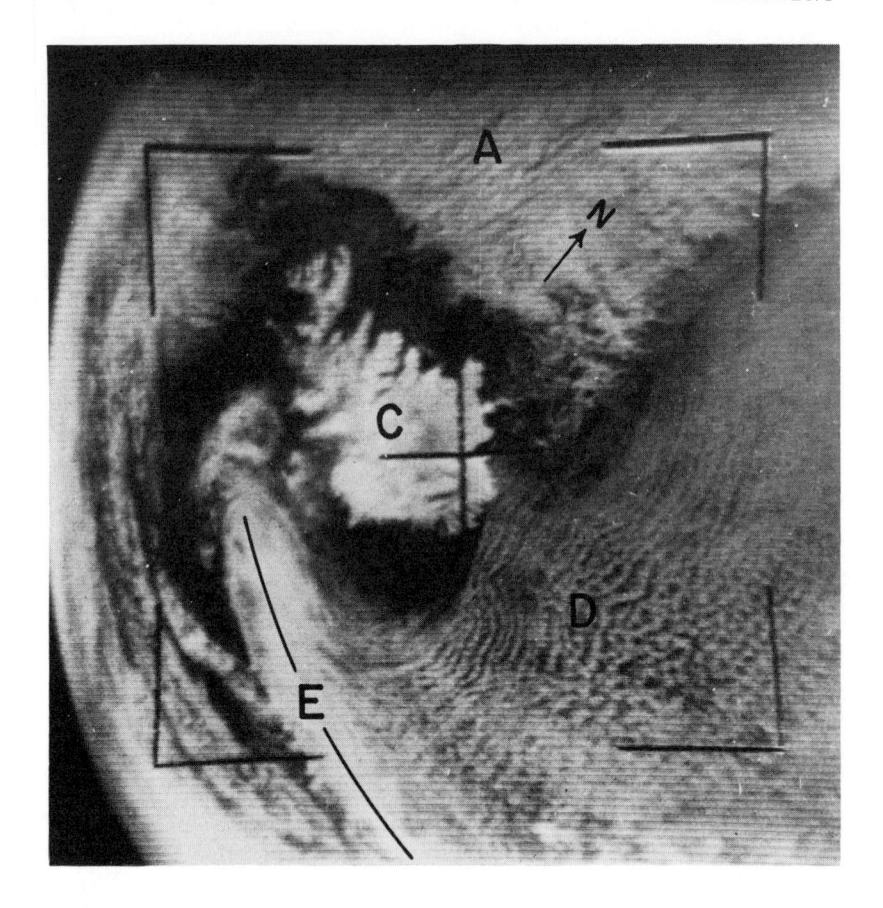

62 BENEFICIAL MODIFICATIONS OF THE MARINE ENVIRONMENT

FIGURE 1 TIROS IX picture (orbit 715/714) March 22, 1965, 1205 GMT. Arrow indicates North. Iceland is the bright area (C) near the center of the picture; the brightness is caused by snow cover. The large gray area (A) north and northwest of Iceland represents mainly an extensive area of ice. (Picture of the Month, *Monthly Weather Review, 93,* 1965.)

In discussing competing climatic change mechanisms, it is useful to divide the basic causes into two types, external and internal. In the case of external causes, we do not expect any feedback between changes in the atmospheric circulation and in the initiating cause. For example, variable solar emission might change the circulation of the atmosphere; but we would not expect changes in the earth's atmospheric circulation to influence events on the sun. Another such example might be that of volcanic eruptions. The large eruption of Krakatoa decreased the solar radiation over the earth for an extended period, but atmospheric circulation does not influence the occurrence of volcanoes.

For internal causes, the situation is completely different. There, the mechanism that starts the change in the general circulation may well be modified by the circulation itself. Moreover, we need to distinguish one "cause" that cannot be entirely ruled out. This is the idea of the random or fortuitous occurrence for which some evidence may exist from model experiments. For example, consider a "numerical model" atmosphere in which eddy coefficients are allowed to distribute quantities with at least some randomness, as is probably true in the real atmosphere. Then, if one starts with identical initial conditions and allows a numerical model of the atmosphere to run on an electronic computer for a large number of days, it is not certain that the atmosphere would evolve in the same way after, say, several years. In this case, the variations might be introduced by round-off errors and other mathematical computational errors, as well as the randomness introduced by the eddy coefficients. Experiments may also be run in a "dishpan," that is, in a rotating cylinder filled with water with temperature gradients introduced to simulate the north–south temperature gradient of the earth's atmosphere. In such "dishpan" experiments, the "atmosphere" (the water medium) will evolve in patterns similar to those found on the earth but, after long periods of time, the atmosphere need not evolve in *exactly* the same way, even though the initial conditions may be almost the same.

The internal-cause type of climatic change is the one mentioned by Fletcher, that caused by the waxing and waning of ice in the Arctic and Antarctic regions. But, if any change in circulation were to occur because of the ice itself, as he has already mentioned, the circulation in the vicinity of the ice would change in such a way as to bring large amounts of warm air to the polar region, and this would modify the ice regime and perhaps reduce its influence. Thus, the effect after time periods of thousands or millions of years would be difficult to predict, and predictions of climate on these scales are entirely speculative. The feedback mechanisms involving the atmosphere, the ice, the circulating oceans, and the earth's surface are very complex. No realistic numerical model exists that can simulate the interplay between these interacting media on the time scale of climatic change. For example, it has been estimated that the circulation in the deep oceans may have periods on the order of thousands of years and that it could influence the climate over long periods of time. However, such ocean circulations have not yet been put into the atmospheric-oceanic models with all the other variables.

A competing idea points to the tropics instead of the Arctic as the seat of the general circulation changes. Bjerknes (1966) suggests that changes in intensity of the equatorial countercurrent, which lies about 3°N of the equator, can influence the circulation over the entire North Pacific and perhaps over the Atlantic too. He did not consider changes on the time scale of ice ages. In his studies, the Arctic was not singled out for special consideration; the tropics were considered the seat of the circulation changes.

Broecker (1968) says

> One cause suggested for the large oscillations in climate that have taken place during the Pleistocene is the change in seasonal contrast produced by variations of the earth's tilt and by the precession of the earth's axes of rotation. Glaciers would presumably prosper during the periods of reduced contrast and diminish during periods of enhanced contrast.

Thus, external astronomical factors would, in this view, control the waxing and waning of ice.

We may also consider the question of intermediate time scales. Fletcher discussed the variations in ice and climate from 1890 to the present. He pointed out that until 1940 the Arctic, at least, was warming, and the ice decreased. After 1940, the trend was reversed. The variation in ice regime may have contributed to the variation of climate, but if the change in ice was the basic cause, why didn't the warming continue? Or, is it more likely that some other basic cause was responsible for both the changes in the ice and in the atmospheric circulation?

In summary then, anomalous amounts of ice doubtless affect the regional weather for relatively short periods. However, the suggestion that polar ice changes are a basic cause of large-scale atmospheric changes over very long periods of time is entirely speculative.

I propose an experiment: If, in some way, the ice could be removed from an extensive coastal region in the Arctic (say, a strip 10 miles wide along the Alaskan Arctic coast or perhaps even along the whole North American Arctic coast), and if this could be done continuously during the winter, the temperature, cloud, and perhaps precipitation climate along the coast would undoubtedly change.

In late spring and summer, the ice might recede more rapidly than normal because of the greater absorption of solar energy by the water in this cleared strip. Doubtless, such a small experiment would not upset the entire world balance of atmospheric circulation and would supply data that would be useful in estimating the effects of changes of ice in the Arctic. But even for this limited experiment, it would be desirable to perform realistic computations before trying it.

REFERENCES

Bjerknes, J., 1966, "A Possible Response of the Atmospheric Hadley Circulation to Equatorial Anomalies of Ocean Temperature," *Tellus, 18,* 820–828.

Broecker, W. S., 1968, "The Cause of Oscillations in Pleistocene Climate" (abstract), *Trans. Am. Geophys. Union, 49,* 113.

Fletcher, J. O. (ed.), 1966, *Proceedings Symposium on the Arctic Heat Budget and Atmospheric Circulation* (RM-5233-NSF), The Rand Corporation, Santa Monica, California.

Mitchell, J. M., 1966, "Stochastic Models of Air-Sea Interaction and Climatic Fluctuations," *Proceedings Symposium on the Arctic Heat Budget and Atmospheric Circulation,* J. O. Fletcher (ed.) (RM-5233-NSF), The Rand Corporation, Santa Monica, California.

"Picture of the Month," 1965, *Monthly Weather Rev., 93,* 368.

Robert D. Gerard and J. Lamar Worzel
LAMONT-DOHERTY GEOLOGICAL OBSERVATORY

Atmospheric Moisture Extraction Over the Ocean

One of our important national goals is the development of new freshwater sources to meet our own growing needs and those of the world. Major efforts and substantial gains have been made in the conversion of waste water and in the desalination of seawater. These efforts all too often take the form of a struggle to overcome the natural environment rather than an effort to use the natural conditions to advantage. Some of the newest large-scale water-conversion systems are now planned as dual-purpose electric power and desalination plants, based on atomic energy. Unfortunately, many of these advances are accompanied by the creation of new problems affecting the environment, such as thermal or brine pollution, atmospheric pollution, and radioactive contamination.

In reviewing the bold engineering advances being made, we are reminded of the manner in which Roman engineers once built their arrow-straight roads across all landscapes, with an apparent contempt for nature. We may ask: Is there an alternative approach to that of applying pure engineering technology to water-resource problems? Can a more complete understanding of the meteorological, oceanographic, and geological conditions lead to methods of water recovery that are more harmonious with the natural environment and that might, indeed, have greater economy?

We have recently proposed a water-recovery scheme that takes these questions into account and that takes optimum advantage of natural environmental conditions. The scheme involves the condensing of atmospheric moisture by the use of cold offshore seawater (Gerard and Worzel, 1967). This paper discusses that proposal and suggests additional uses of the sea to improve desalination plants and to produce power.

Nearly three fourths of the earth's surface is ocean, and 90 percent of the ocean is deeper than 1,000 m. Since the *Challenger* expedition of 1872, we have known that the deep ocean everywhere contains water at temperatures within a few degrees of freezing that originated at the surface in polar regions. In fact, substantially more than half the ocean volume (water deeper than 1,500 m) has a temperature lower than 5°C. The volume of this cold water is 10 times the volume of all land on earth above sea level; it is, therefore, the world's most abundant substance.

Typical temperature–depth curves for various ocean areas are shown in Figure 1. These profiles indicate that throughout the tropics and midlatitudes, temperatures at the 1,000-m depth are between about 5°C and 10°C. Figure 2 shows the principal tropical and midlatitude areas where ocean depths greater than 500 m may be found within 30 km of the shore. Most midocean islands (not shown) are surrounded by similarly steep submarine slopes. These are the areas where, at favorable sites, it may be possible to utilize the cold deep water for atmospheric moisture recovery.

Many of the smaller islands and coastal areas shown in Figure 2 are deficient in fresh water; yet, paradoxically, more than 200 million gallons (760,000 m^3) of fresh water per day, held as vapor in the lower 100 m of air, may sweep across every kilometer of these shorelines.

The conditions necessary to utilize this resource are (a) a coastal location, (b) a position in the regular path of movement of humid, maritime air masses, and (c) offshore ocean depths that provide cold deep-ocean water close to the shore. The last condition does not necessarily require steep offshore slopes; it could also be met by the presence of a submarine canyon, bringing cold deep-ocean water close to shore across an otherwise shallow shelf.

In this scheme, the deep offshore seawater used as a cold source is brought up through a large-diameter pipe and pumped through a condenser array located on shore so as to intercept the flow of the moisture-saturated winds. When cooled, this air condenses much of its moisture, which is then carried away and stored for use as potable water.

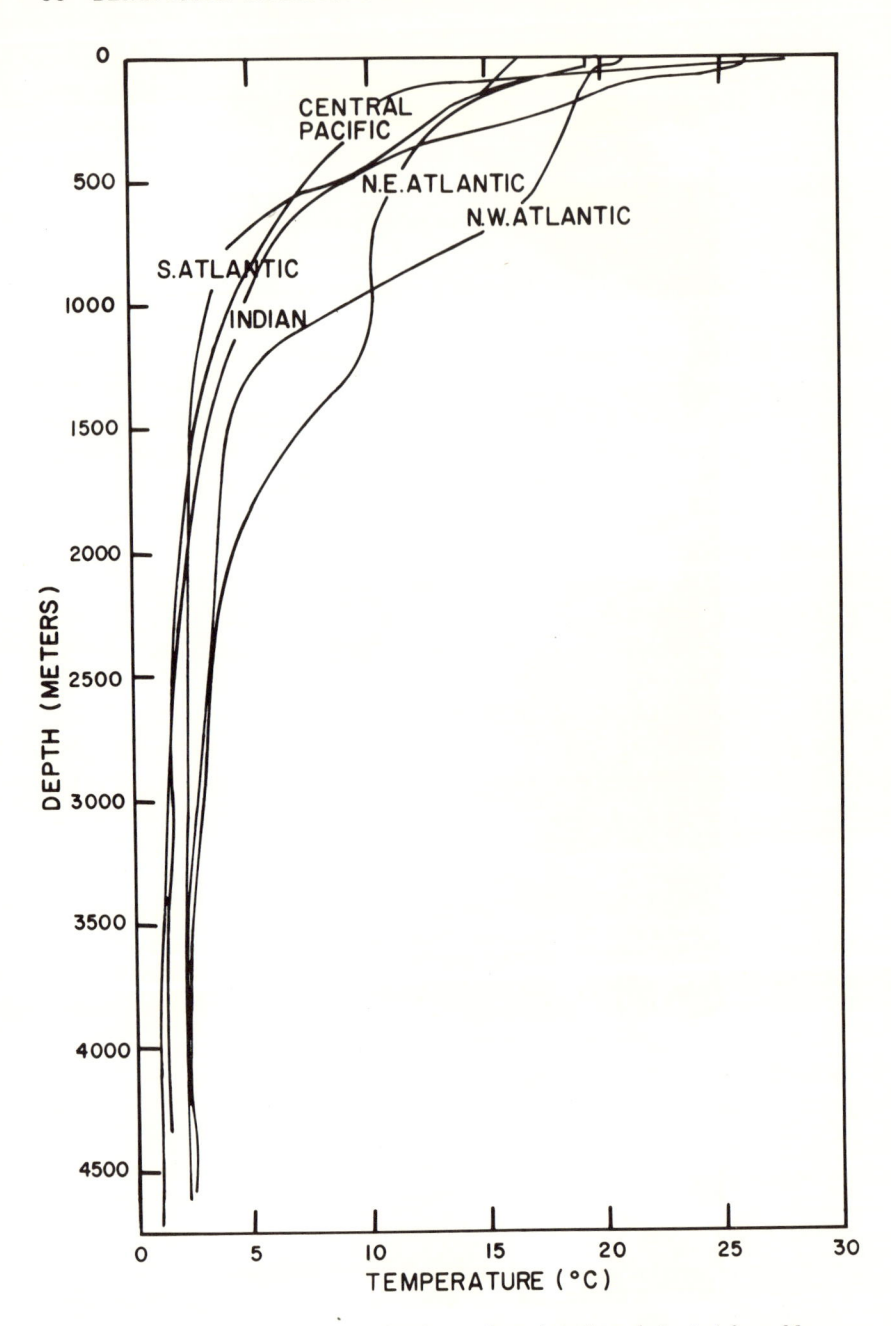

FIGURE 1 Temperature versus depth at selected stations (adapted from McLellan, 1965).

FIGURE 2 Principal tropical and midlatitude areas where seawater of 500-m depth may be found within 30 km of the shore.

The trade-wind areas in particular lend themselves to the use of this system. These zones, situated roughly between latitudes of 10° and 20° in both hemispheres, in the belts between the quasipermanent subtropical high-pressure areas and the equatorial low-pressure zone, cover 31 percent of the ocean surface (Haurwitz and Austin, 1944). Hare (1963) states, "The trade winds have the reputation of being the most reliable constant winds of the globe." These winds are warm and humid, typical characteristics of tropical maritime air masses.

While not notably deficient in rainfall, the trade-wind zones include many oceanic islands that, because they are small and low in elevation, lack conditions that create local rainfall. On many islands, the water deficiency is compounded by high evaporation rates and geological and subsoil conditions that will not support a normal groundwater table.

The Caribbean area, and in particular the Virgin Islands, provides nearly optimum conditions for the operation of this water recovery scheme. St. Croix is typical of many of the smaller islands in the Caribbean. Situated in the trade-wind belt, its rainfall varies from about 65 to 100 cm (25 to 40 in.) annually (U.S. Department of Commerce, 1967); evaporation is roughly equal to rainfall. In spite of unusually favorable groundwater resources on St. Croix, rainfall catchment is the most common water source. In addition, some 200,000 gallons (760 m^3) per day are now produced as a by-product at an alumina-processing plant and sold as potable water. Several small desalination plants are in use, and additional units are planned. The cost of well water delivered by tank truck to certain areas of St. Croix is as high as $14 per 1,000 gallons (3,785 liters). Fortunately, like many other Caribbean islands, St. Croix has a favorable offshore situation for the use of this water-extraction technique.

Figure 3 illustrates the bathymetry of the Virgin Islands area. It can be seen that water of 1,000-m depth is reachable within about 1.5–3 km of the island's north shore. Table 1 contains data from a hydrographic station in the Virgin Island Basin, showing the water temperature in this area to be about 5°C at a depth of 1,000 m.

It is possible to make only rough estimates of size, scale, and cost of the basic elements of our system, since we do not know the exact location, design, and efficiency of the components.

Our requirements begin with a large-diameter pipe, approximately 1.5 km in length, leading outward from shore to the 1,000-m depth. The pipe might require insulation in the upper 300 m of water depth.

Let us assume that a plant producing 1 million gallons (3,790 m^3)

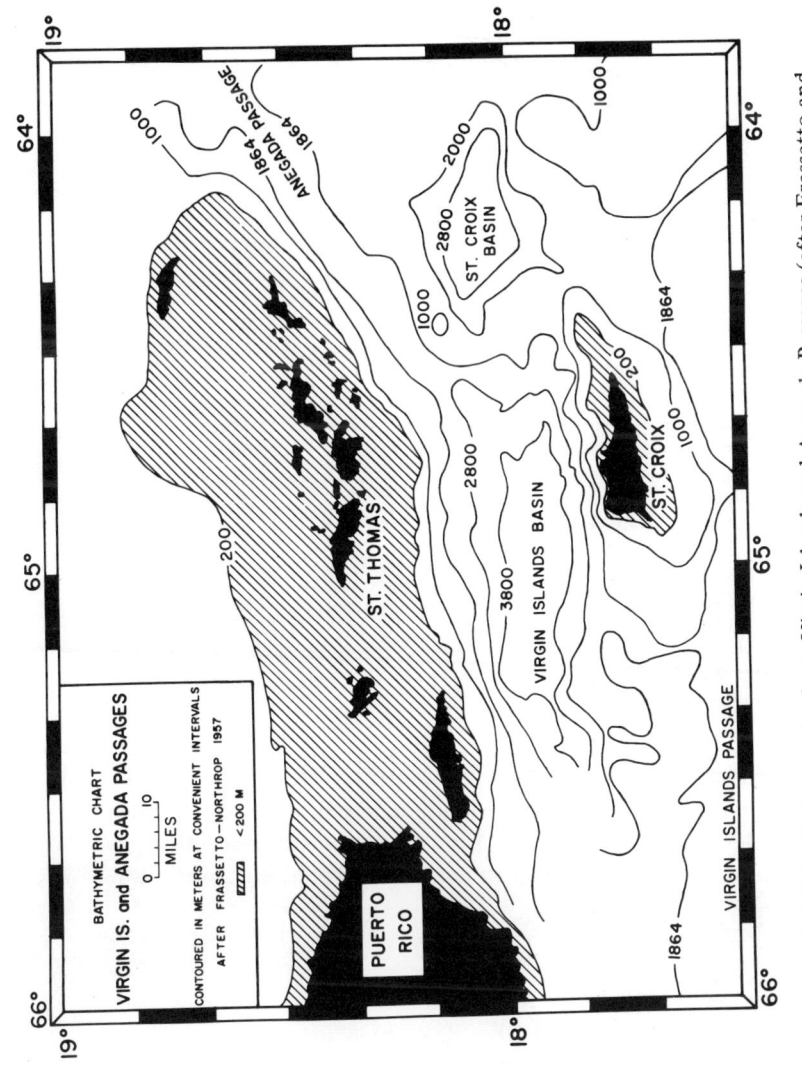

FIGURE 3 Bathymetric chart, Virgin Islands and Anegada Passages (after Frassetto and Northrop, 1957).

TABLE 1 *Atlantis II,* Cruise 14, Hydrographic Station 512, December 9, 1964, in the Virgin Island Basin, 17°57'N, 65°00'W, to depth 4,453 m[a]

Depth (m)	Temp. (°C)	Salinity (per mil)	Phosphorus (μgA/liter)[b] PO₄	Partial	Total	Nitrate N (μgA/liter)[b]	Silicate Si (μgA/liter)[b]
1	26.97	35.162	0.02	0.03	0.15	0.08	1.66
25	26.97	35.158	0.02	0.02	0.11	0.11	2.18
39	26.93	35.152	0.01	0.03	0.13	0.11	1.66
89	25.26	36.957	0.01	0.01	0.12	0.26	0.90
99	24.67	36.938	0.03	0.01	0.08	0.40	0.90
138	22.47	36.928	0.04	0.00	0.45	0.76	0.83
197	19.54	36.686	0.12	0.01	0.16	2.53	1.09
296	17.34	36.381	0.33		0.35	6.15	2.18
395	14.85	35.969	0.76		0.79	11.98	5.31
494	12.55	35.611	1.18		1.17	13.97	9.41
592	9.62	35.189	1.65		1.66	23.01	16.13
691	8.02	35.008	1.87		1.83	26.98	21.06
790	7.02	34.956	1.95		1.95	9.94	22.27
987	5.56	34.957	1.71		1.70	17.31	24.19
1184	4.51	34.970	1.61		1.61	18.58	26.62
1382	4.19	34.978	1.50		1.52	19.66	26.62
1678	4.05	34.989	1.36		1.38	15.37	20.41
1974	4.00	34.994	1.27		0.26	17.76	16.96
2270	3.80	34.985	1.26		1.25	18.85	14.14
2759	3.82	34.992	1.31		1.28	18.44	16.58
3257	3.83	34.983	1.26		1.26	18.13	16.82
3751	3.88	34.981	1.26		1.27	14.01	17.09
4244	3.95	34.985	1.24			17.76	18.62

[a] Adapted from Ketchum *et al.* (1966).
[b] μgA = microgram-atoms.

of fresh water per day is required. Maritime tropical air masses continually sweeping across this island have temperatures that stay within a few degrees of 25°C in all seasons. The factors determining the ratio of cold seawater to condensate water recovered are as follows:

Initial Conditions

Deep water temperature: 5°C (41°F)
Air temperature: 25°C (77°F)
Relative humidity: 75 percent

Assumptions

0.55°C (1°F) rise in deep water due to heat exchange in pipe and pumping heat.

Air Conditions

	Entering	Exiting	Difference
Temperature	25°C (77°)	12°C (54°F)	13°C (23°F)
Humidity	75%	100%	—
Enthalpy	34.8 Btu/lb air	22.6 Btu/lb air	12.2 Btu/lb air heat loss
Water content	0.014 lb/lb air	0.0088 lb/lb air	0.0062 lb/lb air

From the above values, to obtain one pound of fresh water requires

$$\frac{1}{0.0062 \text{ lb } H_2O/\text{lb air}} = 161 \text{ lb air.}$$

Heat loss required for one pound of fresh water becomes

$$161 \text{ lb air} \times 12.2 \text{ Btu/lb air} = 1960 \text{ Btu.}$$

Seawater Temperature

Entering: 5.6°C (42°F); Exiting: 10°C (50°F)

Heat absorbed by seawater coolant:

$$10°C (50°F) - 5.6°C (42°F) = 4.4°C (8°F), \text{ which will absorb 8 Btu.}$$

Thus, the seawater to fresh water ratio is

$$\frac{1960 \text{ lb fresh water}}{8 \text{ lb salt water}} = 245 \text{ lb salt water/lb fresh water.}$$

Thus, a ratio of 245 to 1 of seawater to recovered fresh water appears reasonable. For the proposed freshwater production of one million gallons (3,790 m³) per day, 12 m³ per second of cold seawater is required. This amount could be delivered through a pipe of 2.5-m diameter by a sea-level pump that would consume 270 kW per km of pipe. Additional power would be required to pump seawater to the elevation of the condenser and to overcome heat losses within the system. A fraction of this power could be recovered by the use of a turbine generator as the water flows back to sea level. Figure 4 shows the arrangement of the water-recovery plant schematically. The condenser design is crucial in making the system economically feasible. Dropwise condensation on surfaces plated with noble metal (Erb, 1966) or on Teflon condensa-

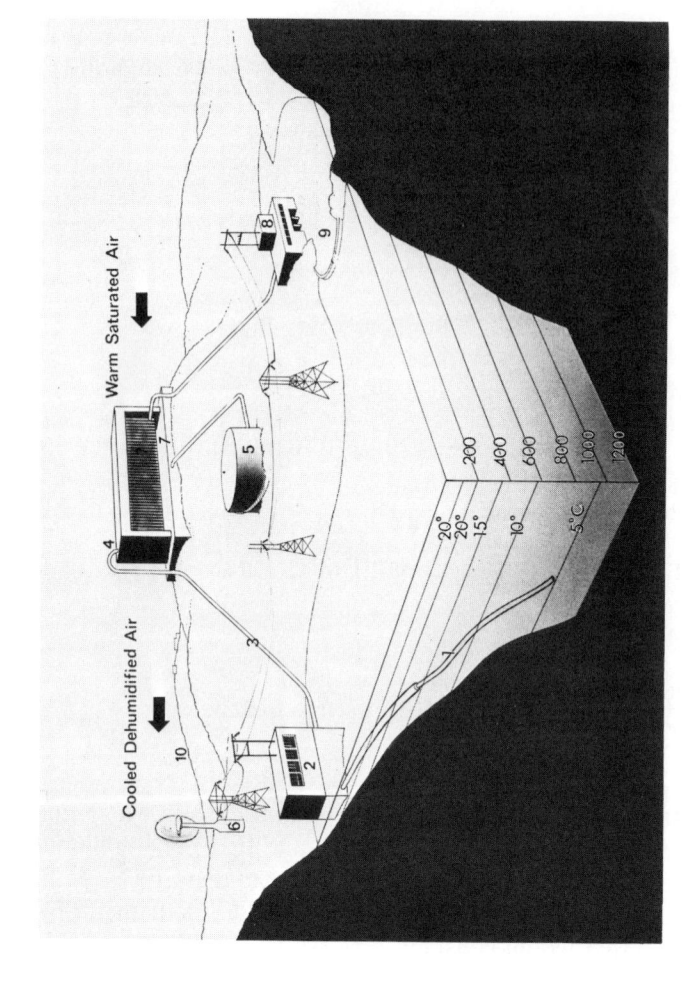

FIGURE 4 Proposed water-recovery plant: (1) large-diameter pipe to deep water; (2) pump station; (3) connecting pipe; (4) condenser; (5) freshwater reservoir; (6) windmill electric generator (for small-scale installations); (7) baffles to direct wind; (8) small turbine to recover water power; (9) lagoon receiving nutrient-rich water for aquiculture; (10) community enjoying cooled, dehumidified air (after Gerard and Worzel, 1967).

tion tubes should greatly improve its efficiency over that of conventional designs.

Wind speeds on St. Croix average 20.4 to 29.7 km per hour (5.5 to 8 m/sec) from northeast to southeast in all months except September and October, when the average wind is 13 to 18.5 km per hour (3.4 to 5.4 m/sec). Calms average about 2 percent of all reported wind readings (U.S. Naval Oceanographic Office and Weather Bureau, 1963).

Taking a wind speed of 24 km per hour and humidity of 70–80 percent, a perfect condenser with a minimum frontal area of about 930 m^2 (10,000 ft^2) would be required. In the worst case, a condenser 30 m high by 1 km in length might be necessary. A forced-air blower system would reduce the required condenser surface area while increasing the overall power requirements.

The strong and persistent winds remind one that windmills provided the principal power for sugarcane processing in the Caribbean islands in the seventeenth and eighteenth centuries. The large Dutch windmills of that period had an output of nearly 100 horsepower. A modern windmill electric generator could produce perhaps four times this amount of power. For a small-scale system in a remote location, the wind might be the most economical power source.

Another possible source of power is the sea. In 1882, D'Arsonval suggested a steam-power generator using the temperature difference between the tropical sea-surface water and deep water. Forty years ago, Georges Claude (1930) actually constructed such a plant in Cuba with an output of 40 kW. It was a scientific success, if not an economic one. More recently, Anderson and Anderson (1965, 1966) suggested a more practical sea thermal-power scheme. They proposed to drive a turbine by boiling propane under suitable pressure at sea-surface temperature and then condense the propane (under pressure) at the deep-water temperature. Their plan for a 10,000-kW plant in the Caribbean calls for pumping 645 million gallons (2.4 million m^3) per day of 5°C deep seawater. They assume that only 14 percent of the generated power would be necessary to run the plant. Cooling water from their plant would be discharged at a temperature only 3 or 4°C higher than its input temperature, suggesting that a vast supply of cool water would be available for a companion plant for the recovery of atmospheric water.

Figure 4 also indicates additional benefits that would accrue, affecting the overall economy of the atmospheric-water recovery system. The depth of 900 m from which seawater would be pumped corresponds closely (in this and many other ocean areas) to the depth of maximum nutrient salt concentration (Table 1). The content of dissolved phosphate and nitrate essential in biological productivity in the ocean is 10 to 20 times greater at this level than at the surface. Ryther (1963) has shown that tropical seas have consistently low productivity, owing to the paucity of nutrients from deeper levels. By contrast, the most productive waters in the world exist in areas where upwelling of nutrients from deeper levels takes place.*

Seawater flowing out from the condenser of our plant would thus be a valuable asset, unlike the harmful brine of a desalination plant. The nutrients contained in 245 million gallons (928,550 m^3) per day of cooling water from a water recovery plant, if discharged into a 60-acre lagoon, could make possible the production of 1,200 tons per year of food fish, or ten times this amount of plankton protein at the second trophic level.

Another important side benefit from this sytem would be that of large-scale "air conditioning." Air with lowered humidity and temperature would be available on the leeward side of the condenser, and advantages in comfort would accrue to those dwellings or communities suitably located.

The proposed system could provide important economies related to scale and location: A small island with a limited technology and foreign exchange is not a likely place to install an atomic-powered desalination plant, but the atmospheric recovery method might readily be used. We believe that a detailed study would confirm the soundness of this plan.

Another way in which natural environmental conditions might be used to improve conventional desalination plants located near a source of cold offshore water is worth mentioning. These plants ordinarily use the input sea-surface water for cooling. A multistage flash-distillation plant might have an operating temperature range of nearly 100°C, based on input water of 24°C and a high temperature of 172°C. The use of cooling water of 10°C would increase the operating range to about 110°C, permitting the use of additional stages and providing proportionately greater output for the same fuel cost. An added advantage would be the reduction of brine-discharge salinity through greater dilution with the discharged coolant.

Other direct uses for cold ocean water can be mentioned: as a cool-

---

*These considerations led Pinchot (1966) to suggest chemical fertilizing and the pumping of deep, nutrient-rich water into circular atoll lagoons in order to raise captive whales. This proposal, which has been termed the "Coral Corral," would take advantage of the baleen whale's high efficiency in turning zooplankton into usable protein.

ing fluid for large-scale air-conditioning systems; and for industrial cooling, where seawater after absorbing the heat load can be discharged at ambient temperature, thus eliminating thermal pollution.

When we consider beneficial modifications of the marine environment—the theme of this symposium—it would be well to stress that the modifications should be mutually beneficial to man and the environment. Development of the needed technology will require the thoughtful contributions of the marine sciences to achieve this goal.

ACKNOWLEDGMENT

Gratitude is expressed to agencies of the U.S. Government, whose generous support through research contracts has made possible this work on ocean problems.

REFERENCES

Air Transport Association of America, 1943, "Meteorological Committee Chart III," reproduced by Tverskoi, P. N. (1965), in *Physics of the Atmosphere*, NASA and NSF Israel Program for Scientific Translations.

Anderson, J. H., and J. H. Anderson, Jr., 1965, "Power from the Sun by Way of the Sea?" *Power, 109*, 63–66.

Anderson, J. H., and J. H. Anderson, Jr., 1966, "Thermal Power from Seawater," *Mech. Eng., 88*, 42–46.

Claude, G., 1930, "Power from the Tropical Seas," *Mech. Eng., 52*, 1039–1044.

Erb, R. A., 1966, "Use of Gold Surfaces to Promote Dropwise Condensation," U.S. Patent No. 3,289,753.

Frassetto, R., and J. Northrop, 1957, "Virgin Islands Bathymetric Survey," *Deep-Sea Res., 4*, 141.

Gerard, R. D., and J. L. Worzel, 1967, "Condensation of Atmospheric Moisture from Tropical Maritime Air Masses as a Freshwater Resource," *Science, 157*, 1300–1302.

Hare, F. K., 1963, *The Restless Atmosphere*, Harper and Row, New York, 106.

Haurwitz, B., and J. M. Austin, 1944, *Climatology*, McGraw-Hill, New York, 48.

Ketchum, B. H., and J. H. Ryther, 1966, "Biological, Chemical and Radiochemical Studies of Marine Plankton," Woods Hole Oceanographic Institution, *Ref. No. 66-18* (unpublished), Station No. 512.

Ketchum, B. H., and J. H. Ryther, 1963, *Machinery's Handbook*, 16th ed., Industrial Press, New York, 1,945.

McLellan, H. J., 1965, *Elements of Physical Oceanography*, Pergamon Press, New York, 28.

Pinchot, G. B., 1966, "Whale Culture—A Proposal," *Perspect. Biol. Med., 10*, 33–43.

Ryther, J. H., 1963, "Geographic Variations in Productivity," in *The Sea—Ideas and Observations on Progress in the Study of the Seas.* Vol. 2, The Composition of Sea-Water, M. N. Hill (ed.), Wiley-Interscience, New York, 358.

Stokhuyzen, F., 1962, *The Dutch Windmill,* C. A. U. van Dishoeck, Bussum, the Netherlands, 34.

U.S. Department of Commerce, ESSA, 1967, "Climatological Data, Puerto Rico and Virgin Islands," *Annual Summary 1966, 12,* 151–156.

U.S. Naval Oceanographic Office and U.S. Department of Commerce, Weather Bureau, 1963, "Atlas of Pilot Charts—Central American Waters and South Atlantic Ocean," *Publication 106.*

DISCUSSION

Helmut E. Landsberg

UNIVERSITY OF MARYLAND

The proposal by Gerard and Worzel (1967) to condense moisture out of the atmosphere is very attractive. It has also the virtue that it is not likely to upset any ecological balances—at least not in the atmosphere.

It is about time that serious thought be given to the exploitation, by engineering methods, of that enormous water reservoir in the air. It constitutes 13,000 km^3 of water. This is, of course, not evenly distributed in space or time.

Actually, the use of atmospheric water vapor through condensation is an ancient art. Curiously enough, but understandably, it seems to have been first practiced by man in the fringe areas of the desert. There, the diurnal temperature range is wide and at night comes close to the dew point. The old shepherds noted that dew would form on rocks that radiated heat to the clear skies. They would collect the dew that dripped off in dew ponds (Midowicz, 1948).

Even in recent times, this has been systematically practiced. Gottmann (1942) reported from the Sahara that a square pyramid of broken limestone, with a baselength of 30 ft, produced about 4 pints of water per day in summer and even a little in winter. Let us remember here that the vapor pressures in the Sahara are between 10 and 15 mb in summer and between 5 and 10 mb in winter. These values are far from the saturation level, but they are not insubstantial.

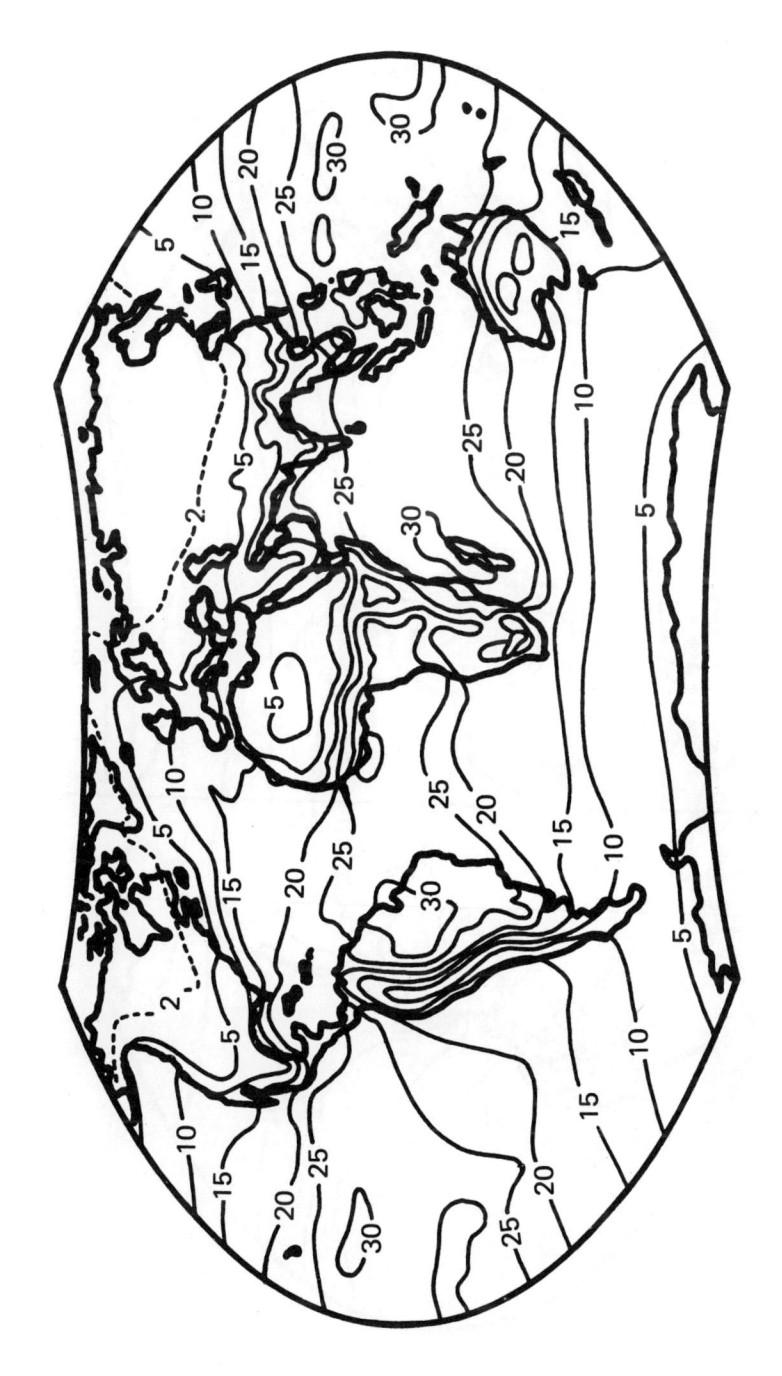

FIGURE 1 Average distribution of the water vapor pressure (in millibars) at the earth's surface in January.

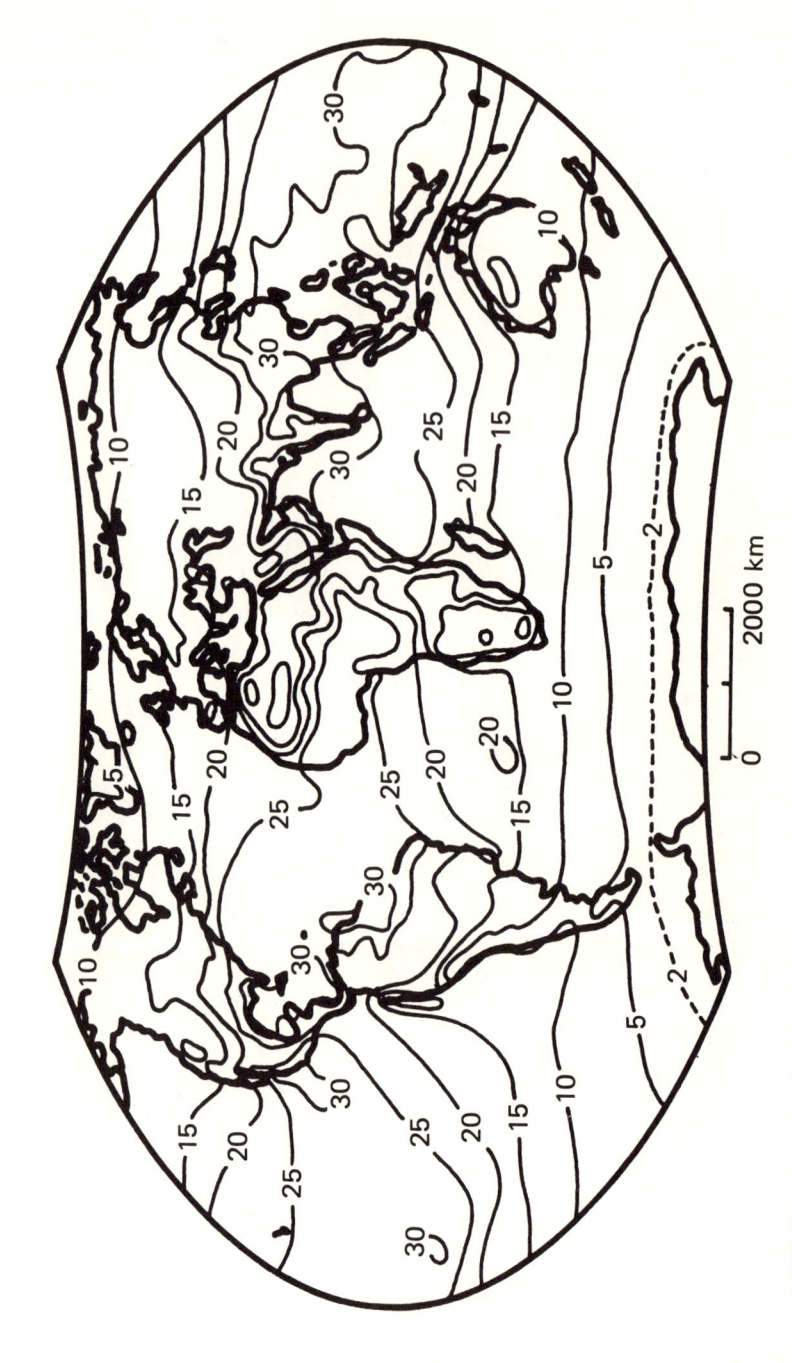

FIGURE 2 Average distribution of the water vapor pressure (in millibars) at the earth's surface in July.

A glance at some mean vapor pressure charts that I produced some years ago (Landsberg, 1964) shows that the area in the Caribbean that Gerard and Worzel discussed is particularly favorable. In winter (Figure 1), the vapor pressures there are between 20 and 25 mb, and in summer (Figure 2), 30 mb or above. As a quick rule of thumb, at the prevailing virtual temperatures, one can say that a numerical value of 73 percent of the vapor pressure in millibars gives the equivalent water content in cubic centimeters per cubic meter of air. That means that at 20 millibars there are about 14 cm^3 of water per cubic meter of air; at 25 mb, there are 18 cm^3/m^3; and at 30 mb, about 22 cm^3/m^3. How much of this can be squeezed out of the air will, of course, depend on the degree of cooling that can be obtained.

If the proposed scheme is further pursued, it might be well to consider also the question of radiative heat exchange. In daytime, one would want to avoid heat loads on the condenser system, perhaps by giving it a high albedo. At night, one could, in many areas, get additional cooling by radiation of heat from the system. In this way, we could secure the joint advantages of the scheme of the old shepherds and the low seawater temperature.

The vapor-pressure charts also show that high values prevail over a vast area of the world, particularly in the intertropical belt. Much of this area is ocean, and many other tropical regions have, because of this vapor supply, adequate rainfall. But there remain enough spots where condensation water may well be important if power to provide cold seawater, as proposed by Gerard and Worzel, is cheaply available. From a climatological standpoint, wind power in its modern guise has much to recommend it in the oceanic trade-wind zones. This aspect has been discussed frequently, and I will refer only to the literature (UNESCO, 1956; World Meteorological Organization, 1954).

REFERENCES

Gerard, R. D., and J. L. Worzel, 1967, "Condensation of Atmospheric Moisture from Tropical Maritime Air Masses as a Freshwater Resource," *Science, 157,* 1300.

Gottmann, J., 1942, "New Facts and Some Reflections on the Sahara," *Geogr. Rev., 32,* 659.

Landsberg, H. E., 1964, "Die Mittlere Wasserdampfverteilung auf der Erde," *Met. Rundsch., 17,* 102.

Midowicz, W., 1948, "An Old Method of Obtaining Water from the Air in Dry Localities," *Meteorol. Mag., 77,* 18.

UNESCO, 1956, "Arid Zone Research," in *Wind and Solar Energy* (Proceedings of the New Delhi Symposium), Paris, 238 pp.

World Meteorological Organization, 1954, "Energy from the Wind," *Technical Note 4, No. 32TP10,* Geneva, 205 pp.

DISCUSSION

Earl G. Droessler*

STATE UNIVERSITY OF NEW YORK

In order to obtain a supply of potable water, Gerard and Worzel (1967) essentially must create cloud conditions inside their gigantic condenser by cooling the inflowing air to 100 percent relative humidity. The experiment would closely approximate the way clouds are naturally formed. Such harmonious working with nature should enhance the chances of success for the scheme.

There is a small body of experimental data on the direct interception of water from fogs and clouds that supports the general thesis of the proposed experiment. It is worthwhile to review the highlights of this work.

Since 1950, systematic field studies have been undertaken showing that direct interception of cloudwater is economically important. The literature contains references to measurements of fog precipitated by trees and a range of papers reporting results using simple man-made implements. For example, Twomey (1957) conducted some experiments on cloud-water interception in Tasmania near the top of Mt. Wellington at an altitude of about 1,400 m. Two rain gauges were used, one collecting normal precipitation, the second collecting the runoff from that portion of a wire-mesh screen directly over the gauge. The results showed that the rain gauge beneath the screen collected ten times as much water as the other gauge. Twomey's was a brief 10-day experiment.

*Now with North Carolina State University.

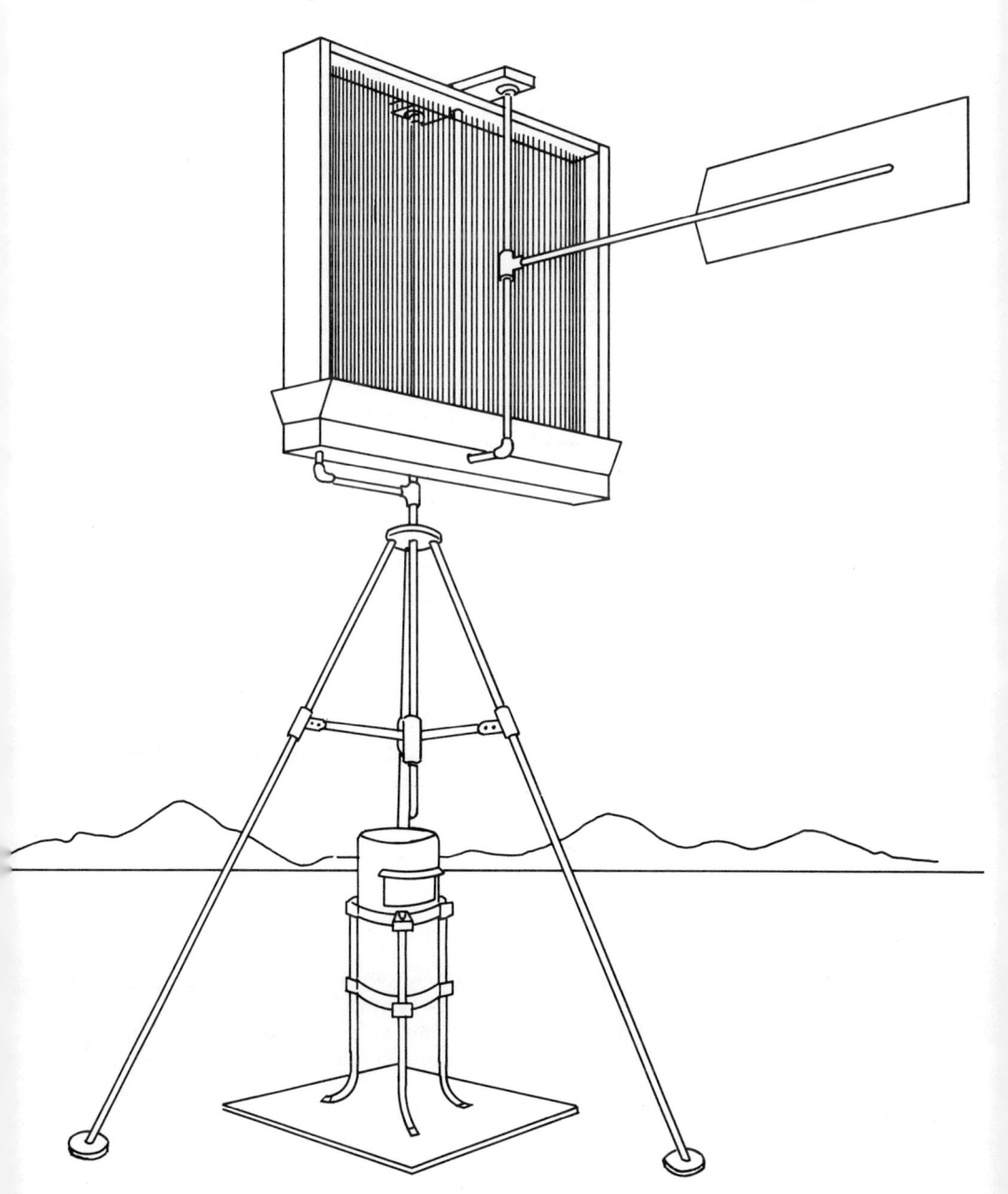

FIGURE 1 One-meter cloud-water interceptor (after Ekern, 1964).

The most impressive results, however, are those of Ekern (1964), who worked for a 3-year period (1955–1958) on the mountain slopes of the island of Lanai, Hawaii, at about 900 m. Ekern reports his results as follows: "Over the three-year period, 1,706 inches of precipitation was collected using an oriented wire harp to intercept the cloud drops while 97 inches was collected in a nearby conventional rain gauge."

Figure 1 illustrates the basic design of the wire harp used by Ekern. Moisture from thin (0.01 in.) vertical copper wires is collected in a trough and piped into a standard rain gauge. The wires are spaced at 1-cm centers in two offset rows.

I made a first calculation of the amount of water that might be gathered with a collector of the Ekern type of the same size as the condenser suggested by Gerard and Worzel: 200 m long by 10 m high. By erecting this device on San Miguel Island off the Los Angeles coast—because we have good data on the occurrence of summertime stratus clouds over the island—a total of approximately 790,000 gallons of water could be extracted from the Southern California stratus during the three months of dry weather (June, July, and August) each year.

This figure does not compare very favorably with the one million gallons per day postulated by Gerard and Worzel. Nonetheless, the method does indicate one other possible means of increasing the freshwater supply for people living in near-arid regions influenced by fogs and stratus clouds of the marine environment. It also emphasizes that the atmosphere contains huge streams of moisture and that man, in reaching up to tap this resource of the air in his quest for more water, is realistic and can be hopeful of success.

REFERENCES

Ekern, Paul C., 1964, "Direct Interception of Cloud-water on Lanaihale, Hawaii," *Soil Sci. Soc. of America Proc.*, 28(3), 419-421.

Gerard, R. D., and J. L. Worzel, 1967, "Condensation of Atmospheric Moisture from Tropical Maritime Air Masses as a Freshwater Resource," *Science, 157*, 1300.

Twomey, S., 1957, "Precipitation by Direct Interception of Cloud-water," *Weather, 12,* 120.

William C. Ackermann
ILLINOIS STATE WATER SURVEY

Water Transfers: Possible De-eutrophication of the Great Lakes

For more than 100 years we have been abusing the Great Lakes with little thought for the consequences. For a long time, the lakes accepted this abuse and appeared to have an infinite capacity for assimilating it, but recently our sins have been showing. Because these bodies are large relative to the inflow and outflow, they have long residence times for their contents, and effects of the sins of three generations now confront us.

In no area of lake studies is this situation so well illustrated as in the consideration of eutrophication, where the critical elements accumulate unseen as dissolved chemical constituents, tasteless, odorless, and nontoxic to man. When certain threshold levels are reached, however, there follows an explosive growth of algae and a damaging chain of events.

In this paper, I shall review the principal physical features of the Great Lakes system and define and describe the eutrophication process. I shall summarize the existing levels of critical nutrients, mention a number of approaches that would slow eutrophication, and finally, I shall discuss a number of water-transfer schemes that have been proposed or that might be worth consideration.

PHYSICAL DIMENSIONS OF THE LAKES

A principal source of information on the physical dimensions of the Great Lakes and the quality of their waters is the report of Special Master Albert B. Maris (1966), dated December 8, 1966, which was based on the extended litigation concerning the Chicago diversion.

The total area of the Great Lakes basin, both land and water, above the head of the St. Lawrence River at the eastern end of Lake Ontario, is approximately 295,000 sq mi. The five Great Lakes—Superior, Michigan, Huron, Erie, and Ontario—together with their connecting rivers and Lake St. Clair, have a surface area of 95,000 sq mi, or about one third of the total basin area.

The international boundary between Canada and the United States passes through all the Great Lakes and their connecting channels, with the exception of Lake Michigan, which is entirely within the United States. Eight states of the United States border on the Great Lakes—Minnesota, Wisconsin, Illinois, Indiana, Michigan, Ohio, Pennsylvania, and New York. The Canadian province of Ontario borders on all of the Great Lakes except Lake Michigan.

Lake Superior, the uppermost and largest of the Great Lakes, has an average elevation of about 602 ft and covers about 31,800 sq mi. Discharge from the lake flows about 70 mi through the St. Mary's River into Lake Huron, which is some 22 ft lower than Lake Superior. During the period 1860 to 1954, the average rate of discharge from Lake Superior was about 75,000 cubic feet per second (cfs).

A gated control structure, placed at the outlet of Lake Superior by the United States and Canada, regulates the discharge from and the level of Lake Superior. The natural supply of water to the lake has been increased through diversions by Canada of water from the Albany River basin, naturally tributary to Hudson Bay, through the Long Lake and Ogoki projects in Ontario (which began in 1939 and 1943, respectively). From 1945 through 1964, the sum of these diversions was an average of about 5,390 cfs.

Lake Michigan has an average surface elevation of about 580 ft and an area of about 22,400 sq mi. Lake Michigan is connected with Lake Huron by the broad and deep Straits of Mackinac. There may be appreciable temporary currents in either direction through the straits, but on the average there is movement of water from Lake Michigan to Lake Huron, the flow ranging between 40,000 and 55,000 cfs. The two lakes stand at virtually the same level.

Water from Lake Michigan and its drainage basin is diverted at Chicago into the valley of the Des Plaines River. On the order of 1,700 cfs is pumped from Lake Michigan by the City of Chicago for domestic and industrial purposes. After use, this water is discharged into the waterways in the Chicago area and flows into the Mississippi River basin. An additional 1,500 cfs is caused to flow directly through the waterways in the Chicago area into the Mississippi River basin. As a result, the water supplied to all of the Great Lakes, except Lake Superior, is reduced in the order of 3,200 cfs.

Lake Huron has a surface area of 23,000 sq mi. The outlet from Lakes Michigan and Huron is through the St. Clair River, Lake St. Clair, and the Detroit River into Lake Erie, which is approximately 8 ft lower than Lake Huron. From 1860 to 1954, the rate of discharge averaged about 189,000 cfs. The distance from the outlet of Lake Huron to Lake Erie is about 84 mi.

Lake Erie has an average surface elevation of about 572 ft and an area of 9,900 sq mi. The natural outlet for the discharge from Lake Erie is the Niagara River, which flows into Lake Ontario, 326 ft lower than Lake Erie. Water from Lake Erie also reaches Lake Ontario by way of the Welland Canal. From 1860 to 1954, the total rate of discharge into Lake Ontario from Lake Erie averaged about 205,000 cfs, about 7,000 cfs of which flows through the Welland Canal.

Lake Ontario has an average elevation of about 246 ft and a surface area of about 7,500 sq mi. Discharge from Lake Ontario flows into the Atlantic Ocean by way of the St. Lawrence River. From 1860 to 1954, the average rate of discharge of Lake Ontario was about 241,000 cfs.

In 1955, the United States and Canada approved a recommendation of the International Joint Commission (IJC) that the level of Lake Ontario be held between 244 and 248 ft above sea level. By its orders, the IJC also created the International Lake Superior Board of Control. The Board is empowered to make and enforce rules in order to maintain the level of Lake Superior as nearly as possible between 602.1 and 603.6 ft.

The mean water levels of the Great Lakes vary from year to year and from month to month. These variations in mean levels correspond to changes in the volume of water in the lakes and result from both natural and man-made causes.

Water is naturally supplied to each of the lakes by rain and snowfall on its surface, by runoff from the adjacent land area of its drainage basin and, to all except Lake Superior, by discharge from the lake above. Water is withdrawn from each lake by evaporation and through its natural outlet river. Water may be either supplied to or removed from the lake through man-made diversions. As the volume of water entering the lake changes, the lake level and the natural outflow adjust continuously, tending to restore the balance between the water entering and the water leaving the lake.

EUTROPHICATION

It seems essential to define and discuss briefly the term "eutrophication," which may not be entirely familiar to everyone.

In general, as a lake ages, it undergoes change, and a natural process of maturation takes place. Precipitation and natural land drainage contribute nutrients that support and enhance the growth of phytoplankton vegetation. The activities of man in altering the landscape by agricultural development and urbanization and the discharge of sewage, industrial wastes, and waste-treatment-plant effluents increase the amounts of nutrient input to a lake. The processes of enrichment and sedimentation that occur naturally are thus accelerated, and the quality of the water may change materially, often at a relatively rapid rate.

The process of enrichment of waters with nutrients is referred to as eutrophication, which, from the Greek, means "well nourished." In the minds of some, little distinction has been given to the differences between the terms eutrophication and pollution. These terms are not synonymous, although it is true that the discharge of waste materials into streams and lakes often adds nutrients that increase the production of both free-floating plants, such as algae, and other forms of planktonic growth.

Since the definition of eutrophication is based on an increase of nutrients, it is evident that measurement of their concentrations is of major significance in assessing the relative state of eutrophication. Many specific nutrients have been studied, including carbon, nitrogen, phosphorus, potassium, silica, trace metals, and vitamins. The most attention has been given to nitrogen and phosphorus because, after carbon, they are required in the greatest amounts for the production of green plants. Nitrogen and phosphorus are supplied chiefly through agricultural runoff containing eroded or leached fertilizers and through urban sewage and storm runoff.

Water Transfers: Possible De-eutrophication of the Great Lakes

An obvious benefit of eutrophication is the increase in the biomass that can be supported in a body of water. Excessive levels, however, are undesirable. Aesthetic values may be lowered because of an increase in algal growths that are a nuisance to those who wish to use the water for recreational purposes. When water supplies are taken from lakes or streams that are undergoing eutrophication, the algae affect treatment-plant operations by clogging filters and may cause undesirable tastes and odors. In the deeper areas of a eutrophic lake, particularly below the thermocline, dissolved-oxygen values may be drastically reduced because of the oxygen demand exerted by the decay of organic materials. Algae grow and die and are succeeded the next year by another crop of algae. As the algae die, they tend to settle to the lake bottom and are attacked by microorganisms that, along with the macroanimal forms in the bottom ooze, exert an oxygen demand. This is what has happened so dramatically in Lake Erie.

The fact that substantial quantities of nutrients are present in lake sediments as well as in the water has been well documented. Therefore, consideration has been given to dredging to remove this concentrated source of nutrients.

Harvesting of weeds, algae, and rough fish is also under investigation, as is flow augmentation that would both dilute the receiving body of water and increase its rate of flushing, thereby decreasing the residence time.

Advanced waste treatment is probably the most effective and practical tool presently available for reducing the level of nutrients in a body of water and thus controlling excessive algae growth.

NUTRIENT LEVELS OF THE LAKES

Numerous federal, state, and university agencies have conducted investigations of water quality in the Great Lakes, and the intensity of study is increasing. To mention only a few agencies does an injustice to the others, but let me say that the water quality information available to me was largely generated by the Great Lakes Research Institute of the University of Michigan and by the Federal Water Pollution Control Administration and its predecessor in the Public Health Service. The most plentiful information available is that for Lake Michigan and Lake Erie.

In general, the waters of Lake Michigan have proved to be of good chemical quality, in that the concentrations of the constituents present are acceptable for both the present and the anticipated future uses of these waters. There has been only subtle deterioration in the quality of the water in the main body of Lake Michigan, as revealed by samplings at stations located 10 mi or more from shore. Sampling within 10 mi of the shore revealed some deterioration in the water quality in areas adjacent to populated sectors of the shoreline. Nitrogen levels were highest in the southwest sector; phosphate was uniformly high except at the surface.

The center of Lake Michigan yielded concentrations of phytoplankton in the 0–500 per ml range; light penetration was generally more than 6 m. The middle of the lake was deep and clear, and contained little organic material. The flora consisted predominantly of diatoms. These are forms that have typified the planktonic algae of Lake Michigan for many years. They are similar to the flora of Lake Superior, which has the highest quality among the Lakes. *Cyclotella,* one of the genera of algae that are not typical of the area but that are common inhabitants of nutrient-enriched waters, sometimes occurred abundantly in Lake Michigan samples and was the genus that predominated in samples from Milwaukee Harbor and from most of the shoreline stations from Milwaukee to Chicago. Total phytoplankton counts averaged about 1,500 per ml within the 10-mi contour and less than 200 per ml at sampling stations beyond the 10-mi contour. Relatively high total phytoplankton counts of 1,000–5,000 per ml were found at many stations along the shoreline. In comparison, Lake Superior averages about 200 per ml at shoreline stations and much less at the center of the lake.

Long-term records of plankton populations in Lake Michigan at Chicago indicate a gradual rise over the past 30 years. Increasing densities of planktonic algae are the primary manifestation of eutrophication. Sawyer (1947), in his work on the Madison, Wisconsin, lakes, gave 0.3 mg/liter of nitrate nitrogen and 0.015 mg/liter of soluble phosphorus as levels at which nuisance conditions may occur.

That the threshold levels of nutrients have been exceeded is evident from the accelerated growths of algae that have been widely noted and reported in the literature and press. Lakes Erie and Michigan have been particularly notable in this respect.

Beeton's review (1961) of nitrogen and phosphorus data collected between 1901 and 1958 for Lake Erie shows that a rapid increase in nutrient concentration is occurring. For example, he reports that nitrates in the western basin increased from 0.10 mg/liter in 1930 to

Water Transfers: Possible De-eutrophication of the Great Lakes   91

0.225 mg/liter in 1942. In the island region, 0.265 mg/liter was found in 1942 and 0.83 mg/liter in 1958. Total phosphorus in the island region increased from 0.014 mg/liter in 1942 to 0.036 mg/liter in 1958.

Ayers and Chandler (1967) have recently authored a monumental report that contains information on nitrates and phosphorus in Lake Michigan. They found that average values of nitrate nitrogen are about 0.10 ppm, although the range was from zero to 0.84. The level was about 0.17 at Chicago and about 0.14 at Milwaukee. With only two exceptions, yearly average nitrate nitrogen values at both Michigan City and Whiting, at the southern end of the lake, have been 0.30 ppm or greater, with highs of 0.52 and 0.50 at Michigan City.

Phosphorus levels in the main body of the lake were in the order of 0.006 ppm—well below the threshold level. However, in the Whiting–Michigan City area, the average was 0.06–0.09 ppm, with levels ranging up to 0.185 ppm.

IMPROVEMENT BY TREATMENT

Before we talk about some ideas for improving the quality of the Great Lakes by introducing water from Alaska or by other means, mention should be made of conventional steps that are available and must be pursued vigorously. These involve the removal from waste streams of critical elements that are now entering the Lakes.

At the Conference on Pollution of the Waters of Lake Michigan and its Tributary Basin, called by Secretary Udall on January 31, 1968, the principal sources of municipal and industrial pollution were identified. At that conference, Mr. A. F. Bartsch of the Federal Water Pollution Control Administration (FWPCA) estimated that 15 million pounds of phosphate are added to Lake Michigan annually. About one third of this derives from agricultural sources and will be very difficult to reduce. However, addition of phosphate from concentrated sources such as municipal and industrial effluents can be greatly reduced by treatment. At the same conference, Dr. Leon W. Weinberger, also of FWPCA, presented information on waste treatment for phosphorus removal. Practical results from very recent research indicate that removal of up to 80 percent of the phosphates is feasible with available technology, at a cost of 5 cents per thousand gallons of treated waste. Obviously, we must pursue this advanced waste treatment without delay to reduce to a minimum the flood of nutrients now being discharged into the Lakes. Research should be accelerated to further reduce these costs and to find practical means for treating agricultural wastes.

NORTH AMERICAN WATER AND POWER ALLIANCE

The most ambitious water scheme was proposed in 1963 by the Ralph M. Parsons Company, and was called the North American Water and Power Alliance (NAWAPA). This plan would collect water from Alaska and northwestern Canada and distribute it to the Canadian prairies, the United States, and northern Mexico.

The total drainage area involved in the collection area is 1.3 million sq mi, with an average annual runoff of 663 million acre-ft. Of this, about 110 million acre-ft, or less than 20 percent, would be withdrawn. NAWAPA would provide for both increased irrigation and electrical power generation. A navigable waterway could be constructed, linking Vancouver on the Pacific Ocean with Lake Superior, and providing irrigation along the way in Canada. The canal would ultimately deliver 48 million acre-ft of water to the Great Lakes system. The cost of the project is estimated at $100 billion over the approximately 30 years required for completion.

There are many fascinating features, but our immediate interest is in considering the effect of adding the proposed 48 million acre-ft of water annually to the Great Lakes system. This flow would enter Lake Superior and add 66,000 cfs to the present average discharge of 73,000 cfs from the lake via the St. Mary's River.

Flow from the St. Clair River above Detroit would be increased 37 percent from the present 176,000 cfs; at Niagara, the flow would be increased 34 percent from the present 194,000 cfs average.

Such flow increases would have apparent advantages for hydropower at Niagara and on the St. Lawrence. However, major difficulties would result for navigation in the St. Mary's River and through the St. Clair–Detroit reach, where current velocities are already higher than desirable. Flushing action would increase in Lake Superior (where it is not needed) and in Lakes Huron, Erie, and Ontario, where it presumably would be desirable. Lake Michigan, which does need the assistance of greater flushing or dilution with high-quality water, would not appear to benefit because of its location as a virtual appendage to the system.

Presumably, the water quality from the Alaskan and Canadian

rivers is good, but no discussion of this is available. However, a cursory review of published data on chemical water quality indicates that the Yukon River at Rampart, Alaska, has total dissolved solids (TDS) of 145 ppm and nitrate nitrogen of 0.20 ppm. The Tanana River at Tanacross, Alaska, has 152 ppm TDS and 0.20 ppm of nitrate nitrogen. The Peace River near the town of Peace River, Alberta, shows 171 ppm TDS and 0.01 ppm nitrate nitrogen. The proposed advantage of diluting the Lakes with these waters, therefore, does not exist according to this limited sample, since the nitrogen levels appear to be higher than those now generally occurring in the Lakes.

My own evaluation of the NAWAPA diversion is that it would be a mixed blessing. Backers of the plan assume that the levels of the Great Lakes are declining and that the area faces a serious water shortage. Both of these assumptions are in error. The levels are cyclical, and various interests periodically suffer from high levels as well as low. There is no water shortage, but there is a serious decline in water quality, and the proposal would only partly solve this while generating serious hydraulic problems.

Perhaps the greatest possible contribution of NAWAPA to water quality would be compensation for a possible diversion out of the Great Lakes watershed of streams that are highly polluted.

WISCONSIN-FOX WATERSHED PROJECT

The Ralph M. Parsons Company has also developed a preliminary proposal to divert 3.6 million acre-ft of water annually from the Wisconsin River, which flows into the Mississippi. By a system of storage reservoirs and improved channels, an average of 5,000 cfs would be directed into the Fox River, Lake Winnebago, and the lower Fox to the Green Bay arm of Lake Michigan. Major hydropower, flood-control, and navigation features would also be included, but our immediate interest is in the effect on water quality in the Great Lakes.

Water so diverted from the Wisconsin River will add inflow to the system and increase discharges by about 3 percent in the St. Clair and less at Niagara. The advocates of the scheme consider the diversion a possible means of compensation for the growing requirements for water from Lake Michigan.

Again, the Parsons report does not discuss the water quality of the Wisconsin River water to be diverted, other than to call it of good quality. Actually, the Wisconsin River is adversely affected by pollution, primarily as the result of inadequately treated municipal wastes that typically receive only primary treatment. Numerous industries—primarily paper mills—also discharge inadequately treated wastes into the Wisconsin. Low dissolved-oxygen levels and other indicators confirm this.

Chemical analyses are available for the Wisconsin River near Necedah and Prairie du Sac, which are just above and below the proposed point of diversion. Nitrate nitrogen during 1961–1964 generally varied from 0.2 to 0.5 ppm, compared to the threshold value of 0.3. Soluble phosphorus varied from 0.02 to 0.05 ppm, consistently above the threshold value of 0.015.

The current emphasis on pollution control and recent improvements in phosphorus removal techniques should make possible a lowering of the levels of these critical nutrients in the Wisconsin River. Under the recently existing conditions, however, the diversion of Wisconsin River water would raise rather than lower the nutrient levels in Lake Michigan and the downstream lakes.

KANKAKEE RIVER DIVERSION

George H. Buck, in his testimony during the Chicago diversion litigation (Report of Special Master, 1966), proposed numerous and varied schemes for the diversion of streams in the Mississippi watershed into Lake Michigan.

One proposed scheme, for diverting water from the lower Kankakee River into Lake Michigan, would require the construction of inlet-control works in the river and a concrete-lined tunnel about 44 mi long running in part through the State of Indiana and terminating in an outlet located offshore in Lake Michigan at a point east of Indiana Harbor, Indiana.

The plan appears to be physically possible from an engineering and geologic point of view. The Kankakee River is known to be of better quality than most streams in the area. In commenting favorably on the plan, Judge Maris (1966) said that it is doubtless better than many streams now flowing into the Great Lakes. Although I can agree with the judge's statement, I think he used the wrong criteria. If proposed diversions into the Lakes are to be considered for quality improvement, they should be better than the average quality currently in the Lakes, not just better than some of the polluted tributary streams.

Chemical analyses are available for the Kankakee River at Wilmington, Illinois, near the proposed point of diversion. During a recent 5-yr sampling period, the nitrate nitrogen levels exceeded the threshold level of 0.30 mg/liter 90 percent of the time. The median concentration was 1.4 mg/liter, nearly five times the threshold.

The soluble phosphorus level in the Kankakee was above the threshold in all samples, with the minimum observed being about twice the threshold. Both nitrate nitrogen and soluble phosphorus were well above the levels now existing in Lake Michigan.

THE GRAND CANAL CONCEPT

The three previously discussed diversion proposals indicate the nature of such plans, but mention should be made of the Grand Replenishment and Northern Development (GRAND) Canal system. This has been proposed by T. W. Kierans (1965), a consulting engineer of Sudbury, Canada. This plan would divert streams flowing into the James Bay arm of Hudson Bay to flow into the Great Lakes. The proposed development would have hydropower and navigation features, would aim to stabilize the levels of the Great Lakes, and would introduce additional new water of good quality.

My examination of the effects of this diversion on nutrient levels in the Great Lakes was extremely cursory because only limited chemical analyses were available. However, all analyses examined for streams flowing into Hudson Bay showed nitrate nitrogen values below the threshold value. No phosphorus analyses were available.

In a very tentative way, I would conclude that in its effect upon eutrophication, the GRAND plan may be the most desirable of the family of such proposals.

DIVERSION AWAY FROM LAKES

Water transfers can be a two-way street, and while we are considering the importation of water to the Great Lakes, we should not overlook the possibilities of diverting away from them the waste products of metropolitan areas. It may well be decided that placing treated effluents still containing certain intractable chemicals into flowing streams where further purification can take place and where the material can more quickly reach the oceans is more desirable than letting them accumulate in the Great Lakes for a century or more.

At the turn of the century, the Chicago River was diverted into the Illinois and Mississippi River systems to avoid returning the wastes of Chicago to its own water supply—Lake Michigan. At the present time (1968), this diversion represents the treated wastes from a population equivalent of approximately 8 million people.

The Chicago diversion has always been viewed with hostility by the other Great Lake states and by Canada, which have collectively been more concerned with quantities than qualities. The effect on lake levels of the Chicago diversion is quite minor—a matter of a few inches.

However, the wisdom of the Chicago action is now becoming more widely recognized and the U.S. Government has agreed that the effluent should not be returned to the lake as was urged by the Lake states. The U.S. Supreme Court has confirmed this in its decision on the Maris report (1966). Perhaps the states will also recognize the soundness of this action. Certainly, it can be said that if Chicago had been discharging its treated wastes into Lake Michigan these past 68 years, the results would not be pleasant to contemplate.

Perhaps this approach of diversion out of the basin can be considered at other places where large population centers are tributary to the Great Lakes, but where a short connecting channel could take their wastes away into an adjoining river system as is done at Chicago. The cities of Milwaukee, Wisconsin; Gary, Indiana; Toledo and Cleveland, Ohio; and Buffalo and Rochester, New York, are examples. These cities have a combined population in excess of 3 million people and a population equivalent waste load of possibly twice this number because of large industrial installations. The quantities of water involved would not be large and the resultant effect on the flow through the lakes and on their levels would be minimal.

INCREASING PRECIPITATION

One of the management techniques that may become feasible in the future is that of increasing precipitation. This is in itself a controversial subject, but there seems to be a consensus that under certain conditions increases of 15 percent are presently possible. Most such reported successes in the United States are for western areas and include orographic effects and generally involve freezing nuclei.

Although I am not aware of any successful precipitation enhancement projects in the Great Lakes area, it is perhaps not unreasonable to assume that, with the major attention now being given to weather modification, increasing the precipitation may become feasible there in the future.

Rough calculations based on increases of 15 percent in mean annual precipitation suggest that the increase in runoff from the water and tributary land surfaces for the system as a whole would be about 63,000 cfs, as compared with the present mean outflow of 232,000 cfs. This increase of 27 percent would not be insignificant and would be welcomed by many of the user interests.

But in the context of this paper we must ask: "What is the quality of precipitation and added precipitation? How does this quality affect the eutrophication process? Is precipitation 'pure as the driven snow'?"

This is not the time to discuss sources of chemical constituents in precipitation beyond mentioning the nucleation process and rainout, and saying that source materials originate from the land, from the sea, and from atmospheric pollution, and that the chemical elements entering rainwater come from both the heights of the stratosphere and from low-level smog.

The chemical composition of precipitation has been studied extensively, and an excellent summary is presented by Junge (1963). However, the elements of nitrogen and phosphorus—of particular interest here—have not been among the most popular elements for investigation. Junge does show a map of average ammonium (NH_4) concentration for the period of July–September 1955 that indicates concentrations of about 0.20 (0.05N) mg/liter for the Great Lakes area. Nitrate (NO_3) concentrations for the same period were about 1.0 mg/liter (0.23N).

In another study by Larson and Hettick (1956), a series of 64 samples of precipitation were collected over the period from October 26, 1953, to August 11, 1954, near Savoy, Illinois. The median sum of ammonia nitrogen (N) was about 0.50 ppm nitrogen, well above the threshold level. The total mineral content of these rainwater samples ranged up to 53.2 mg/liter. No analyses were made for phosphorus.

When these figures are compared with the average total dissolved solids in the Great Lakes—about 150 ppm—it seems safe to assume that added precipitation would not only add water quantitatively, but would somewhat reduce the total mineral content of the water.

A curious situation appears to exist with respect to the nitrogen, in that the nitrogen content of rainwater is higher than that generally existing in the Great Lakes, based on the limited data available. This may be a surprise at first thought, but is not illogical upon reflection. The atmosphere is a principal natural source of the nitrogen needed to support growing aquatic organisms as well as terrestrial plants and animals. Apparently, the processes of harvesting and deposition in the hydrologic system are such that the lake water and inflowing streams are lower in nitrogen than the precipitation.

Artificially increased precipitation presumably would contain lower concentrations of mineral matter than natural precipitation but would certainly add further nutrients. Thus, a very tentative conclusion is that increases in precipitation do not appear attractive as a solution or amelioration of the eutrophication problem.

REDUCING EVAPORATION

In the arid western areas of the United States, it is widely recognized that evaporation from reservoir surfaces is not only an important source of water loss but is a substantial factor in increasing the salinity and total dissolved mineral matter of the water resource. This is a consequence of the fact that, for all practical purposes, evaporation can be considered to be H_2O. Whatever dissolved material is in the stored water remains, and since the volume is reduced, the concentration of dissolved chemicals is increased. The same forces operate in the Great Lakes region, although until now circumstances have not required that this be a matter of concern.

W. J. Roberts (1959) has studied the possibilities of reducing lake evaporation by the application of monomolecular films that retard evaporation but permit the passage of gases, including oxygen and carbon dioxide. Reductions of evaporation on the order of 30 percent are possible in small lakes and ponds, and the method presently approaches practicality for small water bodies when alternative water costs are high. The materials being used, primarily hexadecanol, are nontoxic and are not visible on the surface.

Similar studies have been conducted by others, including the Bureau of Reclamation, on water bodies as large as Lake Mead. The principal problem is one of maintaining the film, which is broken by waves and is drifted by wind action.

An assumption of half the 30 percent reduction in the mean annual evaporation achieved on small ponds would result in an increase of about 32,000 cfs in the outflow of the Great Lakes system. This is a great deal of water—given the assumption.

One should mention that reductions in evaporation from the land surfaces of the drainage area as well as reductions in the evapotranspiration of plants could also result in marked increases of runoff to the lakes. Research is in progress (Jones, 1966) with respect to both soil surfaces and plants, and improved management techniques may become available in these areas sooner than evaporation reductions can be achieved on the lakes themselves. Since the land area in the Great Lakes basin is approximately twice that of the water, it would not be unreasonable to assume that potentially, at least, equal amounts of water could be involved. However, whether improved water quality would result, I am not prepared to say.

Phosphates and nitrates are common elements of chemical fertilizers applied to the land. Decreased evaporation and evapotranspiration would result in increased land runoff with its included nutrients, as would the increased precipitation previously discussed. The resulting concentrations might or might not be greater than those now entering the lakes from streams.

REFLECTIONS

Rainey (1967) has presented a very interesting analysis of the time required for the Great Lakes to purge themselves from a given polluted condition if all further additions of pollution are halted. His analysis is based on simplifying assumptions, but he concluded that it would require 500 yr for Lake Superior to deplete 90 percent of its contamination through natural flow. The corresponding times are 100 yr for Lake Michigan, 6 yr for Erie, and 20 yr for Ontario.

From this we can conclude, particularly for the upper lakes, that if the quality of the Great Lakes becomes seriously impaired, long periods will be required for correction.

My own feeling is that diversions into the lakes should not come from developed areas such as those along the streams in the Great Lakes states. In fact, it would be better to turn away the flows that now drain our metropolitan areas. If water is brought in, it must be of excellent quality and low in nutrients. Precipitation enhancement and evaporation reduction may have possibilities for the future. Certainly, one approach we can use, and must use, is improved treatment of the waste flows now entering the lakes.

REFERENCES

Ayers, J. C., and D. C. Chandler, 1967, "Studies on Environment and Eutrophication of Lake Michigan," *Special Report 30*, Great Lakes Research Division, University of Michigan.
Beeton, A. M., 1961, "Environmental Changes in Lake Erie," *Trans. Am. Fish Soc., 78*, 153–159.
Jones, D. M. A., 1966, "Range of Evapotranspiration in Illinois," *Circular 89*, Illinois State Water Survey.
Junge, C. E., 1963, *Air Chemistry and Radioactivity*, Academic Press, New York.
Kierans, T. W., 1965, "The GRAND Canal Concept," *Eng. J.*, December 1965, 39–42.
Larson, T. E., and I. Hettick, 1956, "Mineral Composition of Rainwater," *Circular 56*, Illinois State Water Survey.
Maris, A. B., 1966, "Report of Special Master, *Wisconsin et al.* v. *Illinois et al.*," *Leg. Intell.*
Rainey, R. H., 1967, "Natural Displacement of Pollution from the Great Lakes," *Science*, 1242–1243.
"Report of Special Master, *Wisconsin et al.* v. *Illinois et al.*," *Leg. Intell.*, December 8, 1966.
Roberts, W. J., 1959, "Reducing Lake Evaporation in the Midwest," *Circular 76*, Illinois State Water Survey.
Sawyer, C. N., 1947, "Fertilization of Lakes by Agricultural and Urban Drainage," *J. N. Eng. Water Works Assoc., 61* (2) 109–127.

DISCUSSION

J. P. Bruce

CANADA CENTRE FOR INLAND WATERS

We should be very grateful to Mr. Ackermann for raising the extremely important question of water quality in connection with interbasin water transfers. Many large-scale proposals have been advanced by the engineering fraternity in the past few years, but these have been concerned almost entirely with water quantities. Little attention has been paid to the likely effects of such transfers on the quality of receiving waters.

Ackermann's paper quite rightly draws attention to the importance of the major nutrients, phosphorus and nitrogen, in hastening the eutrophication of lakes, especially the Great Lakes. These nutrients have resulted in the extensive algal blooms and rapidly advancing cultural eutrophication that has occurred in recent years in Lakes Erie and Ontario. While the extent of algal blooms and degree of eutrophication has been shown to be generally related to the phosphorus loading of lakes (Vollenweider, 1968), and phosphorus is the most controllable of the essential nutrients, in some cases the role of the micronutrients (such as manganese, boron, and iron) may be very important. Unfortunately, we don't know the full story of the causes of harmful algal blooms or of the eutrophication process. This should make us very cautious about the ecological consequences of diversions of water from one natural system to another.

Another important form of modification of the Great Lakes environment might be mentioned. This is the discharge of large quantities of cooling water into the lakes from conventional and nuclear thermal power plants. With most of the continent's prime hydroelectric sites fully exploited, an increasing number of large thermal plants will be built on the shores of water bodies such as the Great Lakes, which can supply large volumes of cooling water. The amounts required are very large; one nuclear plant in the planning stage will take in 4,000 cfs (2.5 *billion* U.S. gallons per day) and discharge it back to the lake 25°F warmer than when taken out. While this is a lot of heat, if it were evenly distributed throughout a lake such as Michigan or Ontario, it would have little effect. But as we learn more about the thermal structure and circulation of the Great Lakes, we are finding that at certain times of the year thermal barriers (the "thermal bar") and circulation barriers (the "coastal jet") will tend to keep the heated water nearshore.

This type of environmental modification may be beneficial ("thermal enrichment") or may be harmful ("thermal pollution"). The surplus energy could conceivably be used to keep ice from harbors and channels, to grow fish, or for other purposes. On the other hand, if not intelligently managed, it will accelerate nuisance blooms of algae. For each 20°F increase in water temperature (between 32°F and 80°F), the growth rate of algae approximately doubles. In short, the choice is ours whether this will be a beneficial or harmful modification of our Great Lakes environment.

Ackermann has referred briefly to the new initiative being taken in Canada to provide the information needed for intelligent water management, with special emphasis on the Great Lakes. A large new water resources research institute, the Canada Centre for Inland Waters, has been established at Burlington, at the western end of Lake Ontario. This Centre will be a major focus for water resources research in Canada and will be large enough to encompass all relevant disciplines, from physics, chemistry, biology, and geology to sociology and economics. It is being sponsored by three federal agencies. The Department of Energy, Mines and Resources has the major coordinating role and will have the largest establishment at the Centre; the Fisheries Research Board will provide the vital biological input; and the Department of National Health and Welfare will be concerned with the bacteriological and health aspects of pollution problems.*

Space is being allocated at the Centre for Canadian university professors and their students to bring the academic view to bear on water problems, and the ship and laboratory facilities will be available to the university community. The private sector, provincial agencies, universities, and federal agencies are represented on an advisory committee. The Centre is temporarily housed in a 25,000-sq-ft trailer complex and a 5-yr, $23 million building program has been approved.

* At the time of publication, water resources components of these three Departments have been combined into a new federal Department of the Environment.

REFERENCE

Vollenweider, R. A., 1968, "Scientific Fundamentals of the Eutrophication of Lakes and Flowing Waters, with Particular Reference to Nitrogen and Phosphorus as Factors in Eutrophication," Organization for Economic Cooperation and Development, Directorate for Scientific Affairs, Paris.

Donald W. Pritchard

CHESAPEAKE BAY INSTITUTE

Modification and Management of Water Flow in Estuaries

Before beginning a discussion of possible beneficial modifications of estuaries, it is desirable first to define what is meant, at least in this paper, by the term "estuary," and to briefly describe the physical properties of such water bodies. In a previous paper (Pritchard, 1967), I defined an estuary as follows:

An estuary is a semi-enclosed coastal body of water which has a free connection with the open sea and within which sea water is measurably diluted with fresh water derived from land drainage.

What are the implications of this definition? First, the requirement that an estuary be semienclosed implies that the circulation pattern is influenced to a considerable degree by lateral boundaries. This boundary restraint is an important feature of an estuary and certainly must be considered in evaluating the possible results of modification and management of water flow in estuaries.

The statement that an estuary is a coastal feature serves to limit to some extent the size of the bodies of water under consideration. The intent here is that the estuary be a part of the coast and not form the coast. Admittedly, the application of this restriction involves somewhat arbitrary decisions, and there are bodies of water, such as the

Baltic Sea and the adjacent Gulf of Bothnia and Gulf of Finland, that would be classified as estuaries by some but not by other students of the field.

The requirement that an estuary have a free connection with the open sea indicates that communication between the ocean and the estuary must be sufficient to freely transmit tidal energy and sea salts.

From the standpoint of geomorphology, estuaries may be grouped into four subdivisions: drowned river valleys, fjord-type estuaries, bar-built estuaries, and estuaries formed by tectonic processes. The drowned-river-valley estuary, also called the coastal-plain estuary, is the most widespread, and probably the most studied, class of estuaries. Discussion here will be limited to coastal-plain estuaries.

Coastal-plain estuaries may be further classified in terms of their degree of vertical and lateral homogeneity. There is a sequence of estuarine circulation patterns associated with the magnitude of the vertical and lateral gradients in the salinity distribution, as discussed by Pritchard (1955, 1967) and by Cameron and Pritchard (1963). I will assume some familiarity on the part of the reader with the major features of this sequence.

Consider a partially mixed coastal-plain estuary. In such an estuary, the salinity distribution will be controlled primarily by the volume rate of inflow of fresh water, the magnitude of the tidal currents, and the physical dimensions of width and depth. These parameters control the relative importance of the various terms in the salt-balance equation and in the equations of motion applicable to the estuary.

The salt-balance equation states that the local time rate of change in salinity is equal to the sum of three terms associated with the longitudinal, lateral, and vertical advective processes and three terms associated with the longitudinal, lateral, and vertical turbulent diffusive processes, or the mixing processes. In most partially mixed estuaries, the primary balance is between the longitudinal advective term and the vertical diffusive term. The tidal currents are the major source of energy for vertical mixing. The vertical salinity gradient depends, however, not only on the amount of energy available for vertical mixing, but also on the rate of inflow of fresh water. The magnitude and vertical variation of the longitudinal nontidal current velocities depend on both the freshwater inflow and the vertical salinity gradient. Hence, while modification of the estuary is most obviously accomplished by modification of one of the parameters (i.e., river inflow, tidal velocity, width, and depth) mentioned above, there is such a complex feedback mechanism that the results of varying one of these parameters are not easily predicted.

The most obvious parameter to control is the freshwater inflow. Gross changes in the character of the estuarine circulation pattern have been accomplished by diversion of fresh water from one watershed to another. An example is Charleston Harbor, located on the estuary of the Cooper River. Some 30 years ago, the Santee–Cooper Project diverted the flow of the Santee River into the Cooper. Prior to this diversion, the amount of fresh water flowing down the Cooper into the estuary was quite small compared to the volume rate of inflow and outflow through any cross section of the estuary due to the tide. The estuary was consequently vertically homogeneous. The characteristic two-layered nontidal flow pattern, with a surface layer flowing seaward and a deeper layer flowing up the estuary, was not present. Instead, there was a small net seaward movement at all depths. The increased freshwater inflow associated with the diversion established a partially mixed estuary having a two-layered flow pattern, with a net nontidal flow directed up the estuary along the bottom. The Charleston Harbor area therefore became a trap for the increased amounts of sediment contributed by the increased freshwater inflow. Dredging required to maintain the channel increased by more than an order of magnitude.

The freshwater inflow to most coastal-plain estuaries along the East Coast of the United States exhibits very large seasonal and year-to-year variations. These variations in river flow produce seasonal variations in both the salinity distribution and the associated circulation patterns in the estuaries. Control of the river discharge through upstream storage impoundments would obviously decrease the seasonal variation of both the salinity and the circulation patterns.

Suppose sufficient upstream storage could be provided that the river discharge to the estuary could be maintained at some nearly constant value. The upstream intrusion of sea-derived salt would then be stabilized, and, except for the tidal variations, salinity at any point in the estuary would be maintained at a nearly constant value. The volume rate of seaward flow in the upper portion of any cross section, and of the landward flow in the deeper portion of that section, would remain approximately constant, and hence the flushing rate of the estuary would not be subject to seasonal fluctuations. Whether such a condition in an estuary would be desirable from an ecological standpoint is a matter best left to the biologists. There are, however, some implications of such a control on the circulation of tributary estuaries that require comment.

An estuarine system such as the upper Chesapeake Bay has a highly dendritic pattern. The only important source of inflow of fresh water to the bay north of the mouth of the Potomac River is the Susquehanna River, which exerts the primary control on salinity distribution in the upper half of the bay. There are a number of tributary estuaries formed by the drowning of the lower valleys of small tributary streams. The circulation pattern, and hence the flushing, of these tributary estuaries is dependent primarily on the processes that control the exchange of water between the tributary estuaries and the bay, rather than on the local supply of fresh water to the tributary estuaries.

Several mechanisms contribute to the exchange of water between Chesapeake Bay proper and the tributary estuaries. In the case of a relatively shallow tributary estuary, such as several of those found along the western shore of the bay, the mechanism which contributes most to the exchange is the variation with time of the salinity of the adjacent bay waters.

In such tributary estuaries as the Magothy River, Back River, Bush River, and Gunpowder River, the amount of local freshwater inflow is insufficient to maintain a significant circulation pattern. At any given time, the water in the tributary estuary is principally composed of water that originated in the Chesapeake Bay. If the salinity of the adjacent bay waters changes with time as a result of seasonal variations in freshwater inflow to the bay, then a salinity difference will be established between the waters in the tributary estuary and those in the bay. If the salinity in the bay is decreasing with time, as occurs in winter and early spring as a result of increasing freshwater inflow, the salinity in the tributary will be higher than the salinity in the bay because the waters in the tributary would have originated in the bay at an earlier time, when the salinity of the bay was higher. Consequently, the fresher, lighter waters of the bay will flow into the tributary in the near-surface layers, and the denser, more saline waters of the tributary will flow out along the bottom as a result of gravitational convection. This mode of flow will continue as long as the salinity of the bay decreases with time, since the salinity in the tributary must lag behind the salinity in the bay.

Conversely, during the period of the year when the salinity in the bay is increasing with time, as during summer and early fall, the bay waters adjacent to the tributary estuary become more saline than the waters in the tributary. In this case the vertical distribution of flow is reversed, to become the more characteristic estuarine pattern of outflow from the tributary in the surface layers and inflow to the tributary along the bottom.

Thus, except during the brief interval of time in late spring and fall when the salinities in the tributaries and the adjacent bay are equal, the flushing rate of the tributary estuaries is controlled by the time rate of change of salinity in the bay. If the discharge of the Susquehanna River were to be controlled to the extent that seasonal changes in salinity in the upper bay were to disappear, then the prime mechanism for the flushing of a number of tributary estuaries would also disappear. Pollution problems within the tributary estuaries would be increased, and the consequent disadvantages might well outweigh any advantages derived from stabilization of the salinity regime in the estuary proper.

Another mechanism contributing to the flushing of tributary estuaries is found in those having channel depths approximately equal to the depths of the adjacent parent estuary. Baltimore Harbor represents such a tributary estuary. A 40-ft-deep channel is maintained in both the harbor and the adjacent Chesapeake Bay. The local inflow of fresh water to Baltimore Harbor via the Patapsco River and several smaller streams is relatively small, so that the vertical mean salinity shows little longitudinal variation between the adjacent bay and the head of the harbor. The vertical variation in salinity is, however, quite different in the harbor as compared to the adjacent Chesapeake Bay waters. In the bay, the balance between the horizontal advective processes (which involve a seaward-directed flow of water of lower salinity in the surface layers and a counterflow of waters of higher salinity in the bottom layer) and the vertical mixing processes produces a relatively strong vertical gradient in salinity. Within the harbor, the advective processes are weakened in comparison with vertical mixing. Consequently, the vertical variation in salinity between the surface and the bottom is less in the harbor than in the adjacent bay waters (Figure 1). The surface waters in the bay are thus lower in salt content than the harbor waters, and gravitational convection requires that the bay waters flow into the harbor at the surface. Near the bottom, however, the bay waters are more saline than the waters in the harbor at the same depths. The bay waters must therefore also flow into the harbor along the bottom. This inflow of

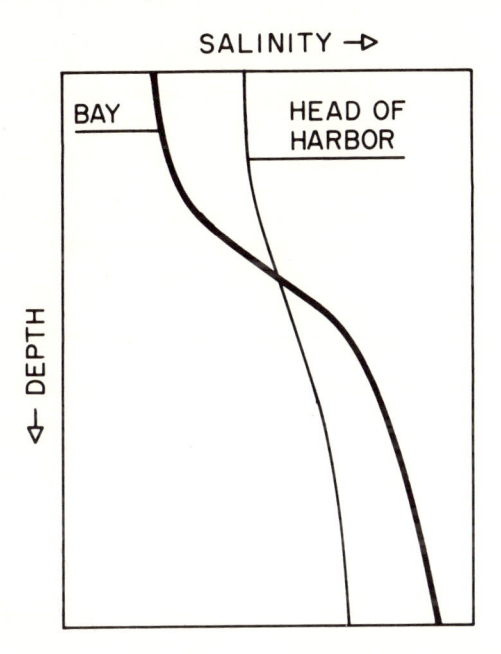

FIGURE 1 Vertical salinity variation in Chesapeake Bay and Baltimore Harbor.

water at the surface and bottom is balanced by an outflow at mid-depth (Figure 2).

The existence of such a mechanism offers the possibility of artificially increasing the flushing rate of embayments that have little or no local freshwater inflow, but that are tributary to a partially mixed estuary and that also have depths greater than the depth of the halocline present in the adjacent parent estuary. One way to increase the flushing rate of such a tributary embayment would be to increase the vertical mixing in the embayment so as to reduce, as much as possible, the vertical gradient in salinity and consequently increase the horizontal difference in salinity between the embayment and the adjacent parent estuary at the surface and at the bottom. (Note that the salinity at mid-depth will in any case be approximately the same in the tributary and in the adjacent parent estuary.) Such an increase in the horizontal salinity gradient at the surface (with the parent estuary having a lower salinity than the tributary embayment) and at the bottom (with the parent estuary having a higher salinity than the tributary embayment) will result in an increased convective inflow in the

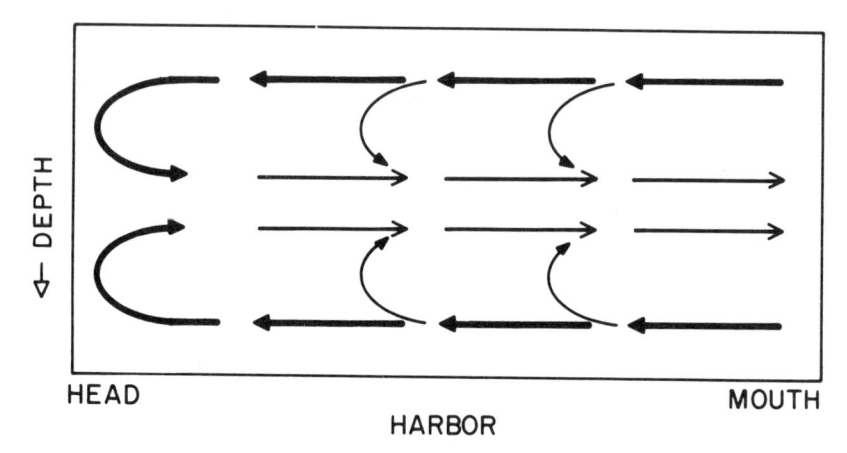

FIGURE 2 Inflow and outflow of Baltimore Harbor.

surface and bottom layers, and hence "renew" the waters of the embayment at an increased rate.

The energy that produces vertical mixing under natural conditions is provided primarily by the tidal currents, and to a lesser extent by winds. It is unlikely that any artificial means could be found to increase the mixing produced by these natural processes. However, it might be possible to increase vertical mixing significantly by the use of air bubbles released from diffusers appropriately located in the tributary embayment. An ascending stream of bubbles acts as an unconfined entrainment pump, setting up a local vertical circulation pattern in which a plume of deep water is carried with the bubble stream into the surface layers. As a result of continuity requirements, an equal amount of water from the surface layer must sink into the bottom layers over a broad area about the ascending plume. Air bubblers have been successfully used around piers, in lakes, and in estuaries to set up vertical convective cells of sufficient strength to inhibit ice formation.

Some experimental studies would have to be undertaken to determine the efficiency of an array of bottom air diffusers in promoting vertical mixing before it would be possible to make a realistic appraisal of the cost of using such a system to promote increased flushing of a tributary embayment.

It should be noted that there are conflicting uses of the estuary, and a modification that is beneficial to one use may be determined to be detrimental to other uses. Marine biologists in Maryland have stated that the estuarine character of the Chesapeake Bay is important to the continued health of both sport and commercial fisheries of the state. It is believed that the highly important striped bass spawning area in the upper bay and the low-salinity barrier to certain oyster pests and diseases require a continued annual discharge of the Susquehanna River at levels close to the present ones. For this reason, Maryland is opposing any significant diversion of fresh water from the Susquehanna watershed, at least until the probable effects of such diversion can be established.

At the present time, the Chesapeake and Delaware Canal is being enlarged from a depth of 27 ft and a width of 300 ft to a depth of 40 ft and a width of 400 ft. This modification will produce obvious benefits to waterborne commerce. At the time the modification was proposed, little or no consideration was given to possible environmental changes in either the Chesapeake Bay or Delaware Bay resulting from the increased potential for exchange of water between these two estuaries. An unpublished study of the 27-ft-deep canal made in 1937 by the Philadelphia District, U.S. Army Corps of Engineers, showed that there was a net flow through the canal from the Chesapeake to the Delaware of approximately 1,000 cfs. Extrapolation of the results of that study, using existing semitheoretical, semiempirical hydraulic relationships, indicates that when the presently approved enlargement is completed, the net flow through the canal from the Chesapeake to the Delaware will be increased to some 2,800 cfs.*

The canal connects to the Chesapeake Bay very near the head of the estuary, and this net discharge through the canal will be mostly fresh water. Hence the enlargement of the canal will produce, based on present estimates, an additional diversion of fresh water from Chesapeake Bay. The volume rate of flow of 2,800 cfs constitutes about 7 percent of the mean annual freshwater inflow into the head of the bay; however, the mean monthly freshwater discharge during summer and early fall is frequently less than 2,800 cfs. The effects of such a diversion on the salinity regime in the upper Chesapeake Bay and, consequently, on the biota, have not been determined. However, this example indicates the complex interrelationships that must be considered when evaluating the balance among the possible beneficial

---

* When this paper was presented in March 1968, this figure was given as 4,000 cfs rather than 2,800 cfs. The new figure represents a refined estimate and has been substituted in the above text in view of the long time interval between the symposium and this publication.

and detrimental changes that would result from a man-made modification of an estuary.

REFERENCES

Cameron, W. M., and D. W. Pritchard, 1963, "Estuaries," in *The Sea—Ideas and Observations on Progress in the Study of the Seas,* Vol. 2, The Composition of Sea-Water, M. N. Hill (ed.), Wiley-Interscience, New York, 306–324.

Pritchard, D. W., 1955, "Estuarine Circulation Patterns," *Proc. Am. Soc. Civil Eng., 81,* 1–11.

Pritchard, D. W., 1967, "What is an Estuary: Physical Viewpoint," in *Estuaries,* American Association for the Advancement of Science, Washington, D.C., 3–5.

Pritchard, D. W., 1967, "Observations of Circulation in Coastal Plain Estuaries," in *Estuaries,* American Association for the Advancement of Science, Washington, D.C., 37–44.

DISCUSSION

L. Eugene Cronin

UNIVERSITY OF MARYLAND

The biologist envies the relative simplicity of the problems Dr. Pritchard has discussed. These are indeed complex, but difficulty increases by at least an order of magnitude when we attempt to take the next step and predict the effects of these changes on biological systems.

Let us consider, for example, the effects of a change in flow into the estuary. One of the fundamental effects is modification of the nutrient input, since the river is the primary source. This, of course, will change the food supply and set new limits on the total populations. Silt loading and distribution will also be different, and benthic organisms may be smothered or their habitat significantly modified. In addition, the osmotic environment will change, and the density and viscosity of the fluid around many organisms will be different as the salinity responds to changes of flow. Physiological effects may range from mild interference with efficiency to death.

The pelagic or mobile species will be affected. River waters supply the chemical materials that influence both the general migration patterns and the seasonal movements of these species. Flows from upstream establish the estuarine circulation patterns, which also have important effects on many animals.

Predators and parasites are also influenced by salinity change, so that the predators and parasites of many species will have a changed distribution.

Thus, the complex biology of individual animals, or species, of various mixtures and, indeed, of entire communities will be under the influence of such a change.

Our present knowledge of these situations is almost entirely qualitative, not quantitative. It will be difficult indeed to make reasonable predictions.

There is, however, evidence of resilience and adaptation in estuarine populations that is generally encouraging to efforts to make beneficial modifications. These species live in a relatively violent aquatic environment, and they have shown excellent ability to adapt. Adaptations can be physiological, structural, behavioral, or even genetic. All of these augur well for the possibilities of constructively changing the environment without necessarily destroying the species present.

It seems to me pertinent to note that the estuary has the greatest potential of all marine environments for manipulated biological production. This is frequently called "aquiculture," though it can take a wide variety of forms. The estuary has these advantages:

It has a high and almost continuous nutrient input from its landward drainage.

There is abundant evidence of its capacity for remarkably high production.

Many of the organisms of the estuary, especially the planktonic copepods and the filter-feeding benthic mollusks, feed low on the food chain and therefore have large quantities of energy available to them.

The estuaries and the tributary rivers are accessible to human manipulation, so modification is more feasible than in the open ocean.

The products of modification might be brought under legal control, which is impractical, and may be impossible, in the open ocean.

Some of the present "wastes" placed in estuaries in growing quantities might be manipulated constructively. Principal among these are the nutrients from domestic wastes and agricultural runoff and the enormous quantities of excess heat from power plants. At present, these nutrients are pollutants or potential pollutants. It is important to consider, however, that each is potentially useful and that each may be manipulated with great benefit and without excessive damage to the environment.

To achieve optimal manipulation of the estuary, we must arrive at effective management policies and put them into widespread practice quickly. The great pressures on estuaries in areas like the eastern megalopolis (sometimes called BOSNYWASH) will not wait, and major areas may be destroyed in the near future.

There is a broad need for increased competent research at every possible level, including the basic mechanisms of estuarine hydrography and biology, and for an assessment of the possibilities for optimal human use of these areas. We must also take advantage of engineering changes, even when they are made without adequate prediction of the principal effects. From such a large-scale "experiment," it is possible to learn much that may have values in many other circumstances.

Finally, we must recognize that the fundamental studies that Dr. Pritchard has reported here, and others like them by physical oceanographers, are of enormous and fundamental importance to any effort to make constructive modifications in the estuarine and coastal environments. They make a valuable contribution to the identification and protection of present values and to the achievement of new and useful possibilities.

DISCUSSION

Joseph M. Caldwell
OFFICE, CHIEF OF ENGINEERS
U.S. ARMY CORPS OF ENGINEERS

Dr. Pritchard's paper and Dr. Cronin's discussion alert us to some of the problems that may arise when man begins to change the regimen of an estuary. It might even appear that the wise thing to do would be to leave our estuaries strictly alone, i.e., to take nothing from them and add nothing to them.

Actually, it would be unrealistic to expect the several million people surrounding such areas as Chesapeake Bay, Galveston Bay, and San Francisco Bay to move away or not to place some demands on the estuary.

Many groups want to use or change an estuary. Their actions might involve one or more of the following:

Diversion of fresh water from the headwaters for other uses
Disposal of human waste products
Recreational boating, fishing, or swimming
Commercial uses (fishing, mining, use for cooling water)
Development of the shore for such purposes as housing, industry, or roads
Protection of the shore from waves and surges
Navigation

These are not evil purposes or activities; the evil comes from ill-conceived or ill-advised uses. Reconciliation of uses is admittedly difficult, but this reconciliation is the job of the engineer, the scientist, and the public official.

Steps (or pressure) to alter an estuary often follow a path somewhat as follows: First, a group or community proposes a use of the estuary that requires altering the estuary in some manner. Then the group petitions the responsible public agency for permission to take the desired action or requests the public agency to design and construct the project. The responsible public agency may then arrange to have a study made in order to provide advice about the effects of the proposed changes and may provide recommendations for or against the proposal. This study is the most important single step, as it is here that the engineering, scientific, and conservation inputs are made and recommendations are received. Finally, the responsible public agency either grants or denies the permit or, sometimes, adopts or drops the project.

Most of these projects have a relatively short fuse. Whatever basic scientific knowledge is missing will probably not be developed in the time frame of the study itself. In other words, the project will be assessed and designed on the basis of scientific and engineering principles that are already available to and generally understood by the professional people studying and designing the project. It is not enough to say that we do not have a sufficient scientific understanding of algae, or fish spawning, or wave action, or hurricane tides, and that we should defer the project until we develop this understanding. Pressure will generally force us to design and to recommend on the basis of currently understood scientific principles.

Thus, it seems that the scientific community has two basic responsibilities with respect to estuary study, or any similar type of study: to see that existing basic knowledge is made available in usable, understandable form to the study agency and to the responsible public agency commissioning the study; and to anticipate the type of quantitative knowledge that will be needed in the future and to see that it is developed before the need is critical.

These two responsibilities, if fulfilled, would represent a kind of scientific Utopia. But we can all hope.